Drawing Botany Home

_____

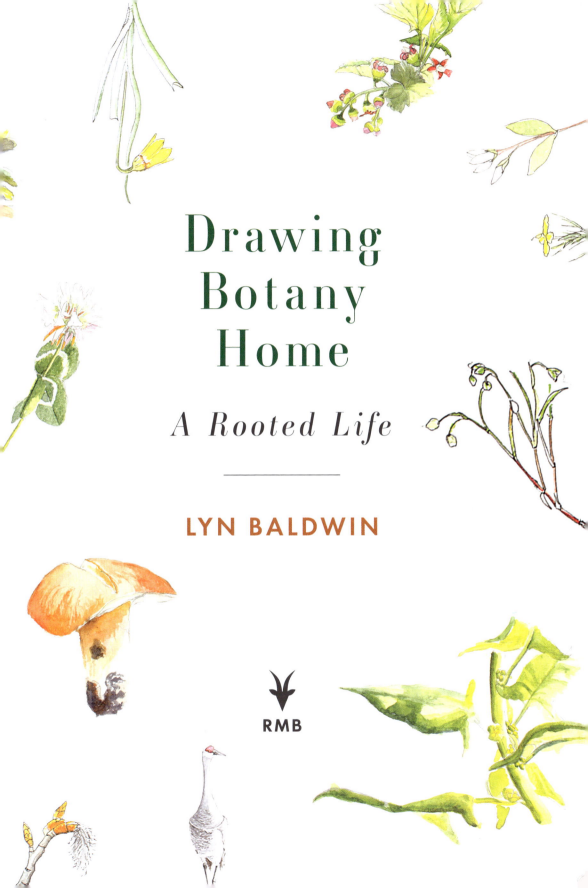

Drawing Botany Home

A Rooted Life

LYN BALDWIN

RMB

For information on purchasing bulk quantities of this book,
or to obtain media excerpts or invite the author to speak at
an event, please visit rmbooks.com and select the "Contact" tab.

RMB | Rocky Mountain Books Ltd.
rmbooks.com
@rmbooks
facebook.com/rmbooks

Cataloguing data available from Library and Archives Canada
ISBN 9781771605922 (softcover)
ISBN 9781771605939 (electronic)

All illustrations are by Lyn Baldwin unless otherwise noted.

Design: Lara Minja, Lime Design

Printed and bound in China

We acknowledge the financial support of the Government of
Canada through the Canada Book Fund and the Canada Council
for the Arts, and of the province of British Columbia through the
British Columbia Arts Council and the Book Publishing Tax Credit.

We would like to also take this opportunity to acknowledge the traditional territories upon which we live and work. In Calgary, Alberta, we acknowledge the Niitsítapi (Blackfoot) and the people of the Treaty 7 region in Southern Alberta, which includes the Siksika, the Piikuni, the Kainai, the Tsuut'ina, and the Stoney Nakoda First Nations, including Chiniki, Bearpaw, and Wesley First Nations. The City of Calgary is also home to Métis Nation of Alberta, Region III. In Victoria, British Columbia, we acknowledge the traditional territories of the Lkwungen (Esquimalt and Songhees), Malahat, Pacheedaht, Scia'new, T'Sou-ke, and W̱SÁNEĆ (Pauquachin, Tsartlip, Tsawout, Tseycum) peoples.

For my mother, Lynne Banks, and my daughter, Maggie Jones.
"Between your two poles, I stretch."

If this is your land, where are your stories?

—J. EDWARD CHAMBERLIN,
quoting a Gitksan Elder during treaty negotiations

Contents

— ACKNOWLEDGEMENTS —

FIRST AND FOREMOST, I am deeply indebted to the First Nations, the Syilx and the Ktunaxa, whose territories helped raise me, and to the Secwépemc People, whose territory helped raise my daughter.

Second, writing, like home, takes practice. I am extraordinarily grateful to the following journals that, in publishing earlier versions of some of these essays, provided important encouragement: *Fourth River* ("Mapping Moss," Issue 0.11, Summer 2021), *Cagibi* ("When Mountains Move," Issue 13, Spring 2021), *Dreamers Creative Writing* ("Dispersal Lessons," Issue 8, March – June 2021), *Terrain* ("Letters to America: Forest Refuge(e)," August 2020; "The Collecting Basket," December 2018; "Community Matters," as "Laura's Collection," April 2015; "Say the Names," Number 29, May 2012), *Cirque* ("Seabound," Volume 11, Number 1, Winter 2020; "Reconciling Botany Pond" as "Finding Home," Volume 4, No. 2, Summer 2013), *Harpy Hybrid Review* ("Collecting the Grip," Issue 1, June 2020), *Hamilton Arts and Letters* ("Carrying Capacity," Issue 13.1, Spring 2020), and *The Goose* ("Form Follows Function," Volume 16, Number 2, February 2018).

Third, no book is written alone. I am indebted to my family, friends and colleagues, who, in ways large and small, supported me in learning plants and place, story and home. For their time in the field, I especially thank Marc Jones, Maggie Jones, Trevor Goward, Curtis Björk, David Baldwin, Laurie Silva, Ron and Marianne Ignace, Mike Rosenthal, Terry McIntosh, Kevin Panewich and Caitlin Quist. For sharing their obsession with the art and science of field journals with me, I thank Libby Mills, Beki Ries-Montgomery, Carol Savonen, Peg Herring, Eleanor Williams Clark, Sandy Bell, Christina Claussen, Margy O'Brien, Sherri York, Hannah Hinchman, Clare Walker Leslie and Briony Penn. I thank Maggie Jones for her patient photography that transformed excerpts of my field journals into images. It's a kind daughter who can tolerate the stress of her mother in the final days of a big project. The essays in this book benefited tremendously from the insights of those who commented on early drafts: Trevor Goward, Nancy Flood, Susie Safford, Jenna Goddard, Dian Henderson, Brian Bouthillier, Eric Bottos, Curtis Björk, Bee Faxon, Angelica Calabrese, Jenn Dean, Jolie Kaytes,

Julie Trimingham, Matt Reudink and Sandra Marquiss. Your feedback echoes throughout this book. In addition, I am deeply indebted to Elizabeth Templeman for not just being one of my most important readers but for recognizing (even before me) the type of stories I was trying to tell. Little in my writing life has been the same since.

I also want to thank Rocky Mountain Books (RMB). One book at a time, RMB's catalogue attests that the land has stories worth learning. I particularly thank Kirsten Craven for her fine and careful editing, Lara Minja for her extraordinary book design that exceeded all my expectations, and Don Gorman for first believing that drawing botany home was a story worth sharing.

Writing this book has convinced me that my botany – in a classroom, on a trail, on a page – needs to serve both those who came before *and* those who will come after. I am deeply indebted to those who came before me. In science, I thank teachers Wilfred Schofield, Gary Bradfield, Paul Bierman, Alicia Daniels, Kerry Woods, Elizabeth Sherman and Edward Flaccus. In writing, I thank teachers Wayne Grady, John Tallmadge and Scott Russell Saunders. I hope this book will carry forward some of the lessons I learned from you. I have benefited deeply from the willingness of Thompson Rivers University (TRU) to support my botany's odd mix of science and writing, drawing and plants. Most of all, TRU has given me a place to learn alongside those who come after – the students who, each year, give me reason to teach. The few whose names appear in these pages stand in for many, many more; I am so grateful to you all. It may not be fair, but *you* are the world's hope.

Finally, every writer has a first reader. I have been blessed with both the careful eyes and rather extraordinary company of a man, Marc Jones, who knows the rhythm of Homer as well as he does the grammar of botany, who risked everything to let me return home. My gratitude is endless.

The Comfort of Buttercups

HERE'S WHAT I REMEMBER.

Brown grass underfoot, sky blue but darkening near the horizon. My little brother and I walking a cow trail mucky with snowmelt. I remember how the trail carried us up and over a gentle hill, wandered us below the pines on the hill's northern side, and then deposited us on the margin of Dry Lake, the shallow pond that formed each spring in the corner of our neighbours' ranch. I remember how the shadows gathered to themselves and then spread from the hollow of the hill. Maybe that was the first time my brother Davey and I found the buttercups in bloom. I'm not sure anymore.

I do remember we had no other place to go. Returning to the house we'd just left wasn't an option. Not right away. I remember wondering how long we'd last before the cold and dark would drive us back. I wondered if Laurie, our older sister, was still inside.

It had started off so well. An early Sunday dinner. Charlie, my mother's third husband, had been jovial, inviting his daughter's common-law husband, Kelly, to eat with us. Kelly, squat and muscular from years of logging the forests of Lincoln County, Montana, never refused food. I remember Kelly slipping in beside Laurie, while Mom brought spaghetti and garlic toast to the table. It had felt okay, the six of us arranged around a table with Charlie and Mom anchoring each end. Kelly staying for dinner was good. Company meant better behaviour. And there was plenty of food. I'd already checked – gauging the size of each serving on the plates against the remaining noodles in the pot. Seconds for all.

I even remember being grateful for a new table. A wooden, hexagonal table, with six rolling chairs that Mom and Charlie had brought back earlier that same day. Charlie was always bringing home something – cheaply bought furniture, or

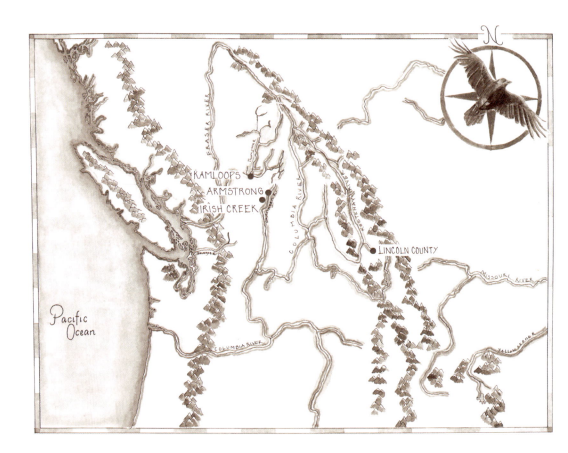

Location Map for Prologue

tools, or other people's castoffs that seemed to have no other purpose than to fill our dirt-floored carport. But the arrival of this table meant I no longer had to share a bench with my brother. Thanks to Charlie, I had my own chair – one that rolled.

Charlie McInturf. A big man, with dark, frizzy hair that bounded upward and a dense, narrow beard that thinned to a point over his too-large belly. In Vietnam, Charlie had traded a chunk of memories for a metal plate in his head. Charlie had left California with his then "old lady" to go back to the land in Montana, only to end up working for the Burlington Northern Railroad. Charlie, whose ex-old lady told my mother the day we moved in, "If someone knocks on the door claiming to be Charlie's kid, go ahead and let them in. Not even Charlie can keep track of them all."

Charlie, whose marriage to my mother rescued my family from the precarious economy of single parenthood in the late '70s before permanently exploding us apart. Within five years, Charlie would be gone; my unwed sister would be mothering her own daughter back home in British Columbia; I would have spent my final year of high school living at my best friend's house; and my little brother would be well on his way to never speaking to our mother again.

But that spring we were all still home, and we had spaghetti and garlic toast. And Mom and Charlie hadn't fought all day.

It happened in a moment – in that proverbial blink of an eye – while I was still relishing the comfort of noodles, while I was still lost in the story I'd been reading before dinner. First, there was the murmur of adults talking. Maybe Charlie was describing the antics of his gandy dancer crew; maybe Kelly was worrying that the mill might stop taking logs if the timber prices kept falling. I don't know, I don't remember. But I do remember the sudden acid in my mother's voice. And I remember Charlie standing, looming really, and with one inarticulate roar, heaving his end of the table at my mother.

There was that moment of impossibility, when the noodles and sauce and garlic toast hung in the air, when the world was devoid of sound, when what was happening was too crazy even for my life. (Who threw tables? Cups, plates, yes, but tables?) And, then, chaos. Plates shattered, utensils clattered, red sauce splattered. Without thought, I grabbed my little brother's hand and ran for the back door. As I led Davey outside, I took one quick look behind, needing a sense of the damage. There, like an island in a sea of debris, sat Kelly. Still in his chair, still forking in spaghetti from the plate he held firm to his chest.

Outside, Davey and I slowed to a walk, chests heaving. I remember we went down to the shores of Dry Lake and mucked about in the narrow fringe, where the grass was verging on green and the buttercups were in bloom. Did I understand the comfort provided – even before I knew their name – in the predictable appearance of these flowers? I don't remember, but I do remember sitting within a sea of buttercup yellow and counting the years until Dave and I could reasonably leave our mother's house: five for me, eight for him. I remember trying to explain to my brother that we just needed to hold on. We didn't have family to run to – we'd only met our biological father a few times, our grandparents even fewer. As the only hippie kids in town, we had few neighbours we could trust.

The one thing I felt I could be sure of within the bounds of my family, other than our mother's indomitable presence, was that Davey and Laurie and I would get older. That someday we would cook our own dinners and eat off tables that stayed comfortably horizontal. I don't know if Davey understood what I was trying to say, but I remember we stayed out there among the buttercups until it was nearly too dark to see.

Time did pass. Dave and Laurie and I did grow up and make homes for ourselves. Yet, in the many years I lived away from my siblings, it has been the tales of our childhood – shared over dinner during my infrequent trips home, lubricated with red wine, performed for an audience of first our spouses and then our kids too – that have most quickly rooted my siblings and me in family.

"Remember when Mom wore that ivory gown with the embroidered flowers to marry Simon in Queen Elizabeth Park in Vancouver? Or the potluck at the Irish Creek cabin in the North Okanagan, when nearly 200 people came, and Mom gave us two gallons of ice cream and all the spoons in the kitchen and told us to go away? Or when our real father showed up in a Mercedes Benz outside our house in Armstrong after Mom and Simon split? Or the Christmas evening the Morlands came over and got stoned with Mom, and then we all went sledding down Okanagan Street in the dark? Or when Mom went off to Montana and married Charlie without telling us, and we moved to Montana three weeks later?"

In my family, as in many, we most often tell the stories that make us laugh, avoiding the ones that don't. Kelly saving his spaghetti is one of our favourites. My siblings and I cast these stories large with the characters of those days – my mother's husbands, the other hippies who colonized the North Okanagan Valley in southern BC in the early '70s, the hard-worn loggers and ranchers we grew to know in northwest Montana. Running, often unsung, through all our stories is the rhythm of a life lived alongside plants. How many times, I wonder, did I flee my family's chaos for the tangled comfort of buttercups? How many of our stories are embedded in the botany of place – woven with riparian willow and dogwood, shaded beneath a ponderosa pine, aching in the stubbled remains of a Douglas fir forest?

ETHNOGRAPHERS AND GEOGRAPHERS tell us that both place and plants matter.[1] The land underfoot, they say, is not just a surface on which our lives unfold but shapes how we think. Our relationship with place, however, is neither simple nor static. Instead, cause and effect circle one another, track back, obscure or entangle. Many of us attach to the places of our childhood, but then their very familiarity allows us to take them for granted. Worse yet, in a globalized and mobile world where millions of us – whether by choice or not – abandon home and country, an allegiance to opportunity or ideas may feel safer than an allegiance to place.

Yet even in our mobile world, few deny the relationship between plants and people; we depend upon them, after all, for food, fodder, pharmaceuticals, fibre and fuel. Yet plants, like place, are easy to miss. Their lives, although complex, are quiet and slow. Botanists warn about the increasing prevalence of what they call "plant blindness."[2] For many of us – including me as a child – plants stand unnoticed, inanimate, a confusing green veil of apparent passivity.

The reality is that plants are never passive. The buttercups my brother and I found 40 years ago were busy – photosynthesizing, growing, expanding, unfolding, fending off predators, wooing pollinators, setting seed. They were also slipping, unbidden, into my life, a stable counterpoint of yellow within a chaotic mass of family.

MY BROTHER AND ME AGAIN. Climbing up the narrow gravel surface of Irish Creek Road. Not on foot, riding in the comfort of his shiny F-150 nearly 40 years later, during the first of our irregular combined family camping trips to Fintry Provincial Park. Outside, it is the height of a hot Okanagan summer – far too late in the season for us to find sagebrush buttercups. Inside Dave's truck, air conditioning raises goosebumps on my bare skin. We're not alone; today our daughters and my brother's new girlfriend, Glenda, ride alongside us. But only Dave and I have memories of this place, riding shotgun as we go.

Memories that, at least for me, overlay this place in a palimpsest of knowing. On the map, Irish Creek Road doesn't look like much: a thin line that switchbacks up through Okanagan Indian Reserve 1 before paralleling its namesake through a narrow valley. But from the moment Dave made the sharp turn onto its surface, it has surged with remembering.

Recognition shudders through my body with each new cue: the lean of muscle against door through the first several switchbacks, the pattern of light and shadow filtering through the canopy above, the smell of rose flowers mixed with dust, the anticipation of the change in slope as we emerge into the hanging valley of Irish Creek.

In the Okanagan aug 19
with the Baldwin-Silva Clan

MARC's FINTRY BIRD LIST   western kingbird, g b heron

Grey catbird
Red-naped Sapsucker
Clarks Nutcracker
Red-tailed Hawk
Flicker
White-winged Crossbill
California Quail
American Robin

· Cedar Waxwing
· Red-breasted Nuthatch
· Merlin
· Osprey
· American Crow
· Ruby-Crowned Kinglet
· Black-billed Magpie

PLANT DOMINANTS

· Sambucus cerulea/mexicana (?)
· Prunus emarginata   J. scopulorum
· P. virginia
· Philadelphus lewisii
· Ceanothus sanguineus
· Populus tremuloides
· Populus balsamifera ssp trichocarpa

Vol. 34: In the Okanagan

e Mt Swanson Overlook
led saskatoon
e both
ial scraped rocks
gon grape
glas - fir
nderosa pine
nnickinnik
niper
nubby penstemon
raea betulifolia
uglas maple
pplssewa
imble berry
milacina racemos
oker's fairy bell
aper birch
uaking aspen
asparilla

VIEW TO THE NORTH - FINTRY BEACH

It's oddly reassuring to holiday in a remembered landscape - a landscape I went to grades 1-6, before our sudden move south. Families change - mostly growing but some losses, too, I think as my siblings and I as well as spouses and kids come and go, from beach to trail in this provincial park ― nestled on deltaic sediments pushed out into the blue water of Okanagan Lake. When Laurie & Glenda drive north to hike Mt Rose, old juxtaposes against new in the never ending weaving of past + present, reality and memory.

FINTRY PACKING HOUSE IN LATE AFTERNOON ―

Irish Creek – the most isolated of our hippie houses. The place I've remembered as the geographic heart of my childhood. An old homestead bought for summer range by a ranching family and rented to my mother and her second husband, Simon. And there it is, a big broad field stretching back to the tree-lined creek. Today, the abundant green canopy above the creek startles me. I remember those cottonwoods best as gray trunks rising up to leafless stems, barely bordered in tawny gold.

Another botanical layer in the palimpsest of place: in the many years I lived away from my family, I carried two photos of my brother. In my favourite, he is sitting, maybe aged 5, strawberry-blonde hair gleaming in the spring sun, knees clutched to his chest, in the middle of the same field that now opens before us. In the photo, taken before the cottonwoods had leafed out, a big open-sided barn is built into the terraced slope on the far side of the creek. A rectangular log cabin sits one terrace higher. Without fail, the outline of these buildings always evoked memories of the rest – the dark root cellar dug into the hillside, the renovated shed that housed our pigs, the broad porch shaded in summer by a sprawling yellow rose.

But right now, in Dave's truck, I am stunned by my realization that not only can I *name* Irish Creek's plants, but that, in this place, each name feels weighted with an importance I don't fully understand. In all the years I spent yearning for this place, I never thought to populate it with the specific names of trees. It was, and always has been, in my memory, *Irish Creek*. Down low on the first switchbacks, ponderosa pine and Douglas fir – interspersed with openings of blue-bunch wheatgrass and sagebrush; mid-slope, a more mesic forest of paper birch and western red cedar above a layer of rose and snowberry. Up here, in Irish Creek's valley, paper birch lines the fields and black cottonwood, the creek. Why, I wonder, were these names never a part of my childhood?

It wasn't from a lack of intimacy. As a child in both BC and Montana, I would have given much for a little distance from the rooted lives of plants. Say, a house with a white picket fence, or a family supported less by garden produce and more by regular paycheques. In 1978, our abrupt move from the North Okanagan Valley to northwest Montana may have changed the country in which we lived, but it did little to influence *how* we lived.

Often, our houses lacked running water, sometimes electricity. My mother grew impossibly large gardens, whose weeding swallowed summer days, and the need to replenish our woodpile gobbled most autumn weekends. Except for one year near the end of our time in BC, we occupied houses at the end of long driveways that drifted with snow in January and sank into mud in March. The *Last Whole Earth Catalogue*, with its bold cover depicting a crescent-shaped Earth emerging from black space, was the closest we ever had to a family Bible.

As a child, I had no interest in plants. Certainly, I had no pride in the skills I'd learned as a hippie kid living off the land – how to sharpen and grease a chainsaw, clean seeds from an ounce of pot, run a rototiller, or drive an old International four-wheel drive along logging roads in search of moss for the hanging baskets my mother sold at local farmers' markets. Did I object to the intimacy we had with plants, or just our poverty? Or to the isolation I felt from the straight kids with their clean clothes and washed hair? In my mind, all three were inextricably linked.

Now I know I wasn't alone. Dispersed across not just BC and Montana but all of North America, there was a crowd of us kids who had gone back-to-the-land, whether we wanted to or not. One researcher estimates that, during the 1970s, nearly 100,000 people went back-to-the-land in Canada, while nearly 900,000 did the same south of the border.[3] Today, 40-plus years later, some of us kids are still angry, writing in *Salon* that it was never a good idea to let your kids smoke pot.[4] Some are still immersed in the lifestyle; still others have found the funny in our history. Together, we are the Adult Children of Hippies, the ACOH.

Say it loud, say it proud. I didn't know that the minority I belonged to had a name, let alone an acronym, until a friend sent me Willow Yamauchi's book, *Adult Child of Hippies*. It's the only book I've ever read aloud to my older sister. On the phone, all 144 couplets: "You know that you're an Adult Child of Hippies...if your ability to split kindling in the dark earned you heavy praise from your family." Or, "You know you are an Adult Child of Hippies...if you could 'sex' pot plants by the time you were five."[5] That night, as laughter ricocheted between my sister and me, my daughter, Maggie, asked, "Daddy, what's a hippie?"

Might I, as a kid, have been more supportive if I had known that going back-to-the-land in North America has a history spanning at least several centuries?[6] That British Columbia had been a destination for those seeking to establish utopian communities as far back as the 1860s?[7] For his experiment living at Walden Pond, many consider Henry David Thoreau the first of the back-to-the-landers. Today, the parallels between Thoreau's retreat to Walden and the 1970s back-to-the-land movement are clear. Both were attempts to escape "the growing industrialization, urbanization and commercialism" of their times.[8] Both forsook household amenities. It took only one visit to my grandparents' suburban house in Coquitlam, BC, for me to understand how little – other than an abundance of gardens – my childhood matched my mother's. Hers, I quickly counted, had several comforts I longed for: electricity, running water and flush toilets.

I first read *Walden* in university, not long after I left my mother's house. When I find my copy still on my bookshelf, it contains few marginalia, but in the "Beans" chapter, there's one sentence underlined: "Why concern ourselves so much about our beans for seed, and not be concerned at all about a new generation of men?"[9] I wish I could remember why I marked *that* sentence. Was I indignant, having learned first-hand the importance of *knowing* beans if your dinner depended upon them? Or did Thoreau's familiarity with beans finally legitimize the life I'd known as a child?

Thoreau spent two years living in the small house he built at Walden Pond. My living off the land ended the day I moved into my university dormitory in Bennington, Vermont. Geographers explain that it is our *experience* of space – the memories we carry, the people who are beside us, the intentions we have – that transforms *space* into *place*. My dorm room wasn't fancy: maybe 12-foot by 12-foot, with white walls and a yellow-orange carpet worn flat. But it was built for two. To my roommate, a nice girl from a small city in Massachusetts, I think our shared dorm room was, at best, a modest space not far from home. To me, it was the first place I'd lived where I could summon heat without first splitting wood into kindling.

In retrospect, it seems ironic that I learned the names of plants native to Thoreau's home before I learned those of my own. In Irish Creek, Davey and I played beneath Douglas fir, western red cedar and paper birch, but our fairy tales had roots in European, not North American, forests. My mother cultivated plants inside and out – replacing bedroom walls with plant shelving when she ran out of space in her greenhouse – but the plants my family grew were those brought to North America by European settlers. Sculpted by their long interaction with humans, most crops (like wheat, peas and carrots) are annuals, overwintering above the soil as seeds or tubers. All plants tell stories: those we plant in fields are like books, packaged and discrete – portable.

My family may have gone back-to-the-land in the 1970s, but we lived like our crops: dependent upon the soil but never relinquishing the possibility of our next move. How much would be different, I wonder, if I had known there were stories that had, since time immemorial, rooted people in place with the plants growing outside our garden?

FROM MY SEAT IN DAVE'S TRUCK, it's unclear what remains of the Irish Creek homestead, but my brother doesn't hesitate. Bumping into the ruts of its long driveway, I am a small girl again, sitting on the lap of our artist friend, Bob Masse, steering his green panel van toward the creek. But the closer we come to the creek, the more uncomfortable I feel. It's one thing to populate a remembered forest with the Latin names of trees; it's another to arrive unannounced at what is clearly someone else's home.

And, in this part of BC, I'd be happier if we weren't arriving in a big, shiny truck with Alberta plates. When we thud, thud across the bridge timbers, I'm disoriented. The geography is all wrong. The driveway goes above, rather than below, the log cabin. No, it's not the log cabin, it's a new house, built where our pig shed once stood.

No one's home. Dave turns the truck around, even as I crane for a glimpse of the old root cellar – the one whose roof boards we scavenged for firewood when Simon had left. As we drive back out to the road, Dave's girlfriend, Glenda, is asking questions about our life here. It's clear from my brother's responses that he still counts in the angry column of ACOH. I'm not surprised; Dave and my mother have not spoken in decades. But I can find little trace in his words of the deep affection – for lack of a better word, the reverence – for Irish Creek that is echoing through my body. Its botany is perfect: a mesic forest, riparian habitat, open meadows.

It's easy, driving back downhill, to imagine the pull of this place in the 1970s. To imagine how the comforts of running water and electricity might have paled in comparison with the chance to live beneath these trees, within this intact ecosystem. For all my childhood resistance, is the pull I feel today, deep in my gut, proof that the hippie back-to-the-land experiment worked? If so, why did I always yearn for this forest and not the one I grew to know in Montana?

PLACE, THEY SAY, IS SPACE WITH MEANING. I called Irish Creek home smack dab in the middle of my middle childhood, a time when many believe we are particularly open to the world.[10] Living in Montana, not only was I older and more contained but I was made strange by both my family's lifestyle choices *and* our citizenship. Within the two landscapes of my childhood, the specific botany might have been largely the same, but my sense of belonging never was.

My brother and sister returned to BC less than five years after my mother married Charlie McInturf; it took me more than a quarter-century to find a way back. In 2004, when a job offer gave me – along with my husband and daughter and our two dogs – reason to move to Kamloops, BC, in the South Thompson Valley, just northwest of Irish Creek, I never questioned the cost of yet one more move. Mobility, I assumed, was a part of being a professional botanist, and after living for 27 years in the United States, this move would allow me to return to southern BC, the land from which I'd long been exiled. Finally, I would be a Canadian living in Canada, in the mosaic of forest and grassland I remembered as home.

In late 2004, I thought I was returning home to teach botany and ecology. I carried with me the tools of a plant ecologist: hand lens, identification guides, statistical software packages. I knew the grammar of Linnaean taxonomy and could decipher a plant community for its ecological processes. I'd even learned to draw plants to help remember their names. But little did I understand that home is never just a spot on a map. Little did I expect how close my homecoming would come to failing.

But it did.

Nothing was as I expected – the job, the place, the country. Discouraged and homesick, I did what I'd always done as a hippie kid when things got hard: I ran outside.

At the time, drawing plants had already blossomed into volumes 1–9 of an illustrated journal practice, but these journals were a hobby, part of what gave me comfort when I ran out the back door. I never went far – rarely more than a day's drive from our new home – but I went with my field journal in hand. Within this range, the plants of this landscape, infused with both memory and meaning, began to act as a refracting mirror, challenging my assumptions about the science I called my own. My botany was not, as I'd always assumed, a placeless science. My botany had always been rooted in both science and art, long supported by those allied with place, nourished both by the ground underfoot and nutrients washed in from afar. Botany, I also came to recognize, was never neutral; the same discipline that had given me both professional and financial stability had always been a colonial science. A science that long ago stripped others of community and that, even today, continued to privilege the global over the local.

Thus, even as the rhythm of drawing plants rooted my botany, it questioned what *world* I wanted it to support. I came home with the world view of a scientist, convinced of science's capacity to conserve the species and ecosystems I loved. Yet it became clear that counting plants was not enough. Thirty years ago, naturalist Robert Michael Pyle warned that our society's increasing *extinction of experience* with the living world might be its greatest threat.[11] In North America, direct knowledge of organisms in their environments (especially plants) has been in severe decline for more than 50 years. How can we hope to save what we don't know?

It was also clear, however, that in the scientific practice of botany, there was little room for botanists' *experience* with plants. In science, we are as separate from the objects of our study as North America once was for most Europeans. Like other scientists, I'd become convinced that if we really wanted to help transform our society's world view, if we really aspired to "living within, as opposed to outside or above nature,"[12] we needed to make room for the disruptive, transformative power of art alongside our science. In changing behaviour, art – both its creative processes and its products – can serve as a catalyst, transforming both its practitioners and the world.

I was no artist. But if my field journals had originally arisen from the same dictates that commanded early European naturalists to make "on-the-spot" records, I knew these hand-bound, and inevitably battered, journals had somehow been making space for me to learn not just *about* but *from* plants. I began to wonder if field journals – with their inherent mix of image and text, art and science, sometimes even poetry with numbers – could be tools with which to reinvigorate our *experience* of place and plants.[13]

In one of the scariest applications I've ever made, I proposed to use my field journal to learn from plants in place. I flogged exhibit proposals to art galleries and submitted illustrated essays culled from my field journal. Little did I know that the challenge in taking plants seriously – as independent beings with lessons to teach – is that we nearly always end up reconsidering ourselves. Drawing botany home resulted in unexpected metaphors, new ways of thinking, that allowed me to acknowledge both the dark horror of colonialism embedded in botany's traditions and the hard bits of my own family history – childhood poverty, a sharp-tongued mother, unwanted stepfathers, a traumatic fire. Most importantly, drawing botany home forever reimagined the intent of my botany.

Returning to BC in 2004, I was haunted by my own mobility, but I had not yet learned to fear the Anthropocene, the Age of Humans. I had not yet lived in a world that raged with wildfire, whose atmospheric rivers washed away highways, whose forests no longer carried the seeds for tomorrow, whose human mobility initiated a global pandemic. If I thought about the relationship *between* plants and people, it was only to worry about what we humans *did* to plants. I did not yet understand how so much of the Anthropocene rests on the stories we humans choose to tell *with* plants.

Some of this is just the numbers: by biomass, plants make up nearly 80 per cent of terrestrial life and, in the Anthropocene, we humans are the largest geologic and evolutionary force in the world.[14] What humans do with plants shapes the world. But for too many of us, our plant blindness means we never question the nature of our relationship with plants. Blind to plants, few of us consider how our grocery store vegetables represent a broken link between people and place, how most of us subsist on plants made portable by domestication, grown in soils we will never walk, harvested by hands we will never thank. Blind to plants, few of us consider how our relentless hunger for crops, timber and paper has stripped our world of nearly one-half of its trees.[15] Blind to plants, few of us sense the complicated tangles of colonialism and globalism that haunt the plants we use to flavour our cooking, to sweeten our drinks or colour our gardens. Blind to plants, few of us know that the survival of the world's insects – 40 per cent of which have declined in the last decade[16] – depends upon our willingness to make space for the plants they need. Blind to plants, we miss the chance to learn from them, to see their rooted bodies as a strange and wondrous alternative to the human mobility that threatens so many of the world's ecosystems and places.

Today, this is what I believe: to meet the ecological and social challenges of the Anthropocene, we must all re-story our relationship with plants and the ecosystems that support them. Cultivating care – between plants and people, people and place – is both essential and necessary. In *Drawing Botany Home*, the lessons I have learned from plants, field journal in hand, haven't always come in order; often I understood the significance of a moment only much later. Thus, the essays that follow are not strictly chronological. When my daughter, Maggie, appears in these essays, her age helps mark time. Sometimes I am alone; other times I travel in the company of my students, my family or our dogs – first Perra and Shasta, and then, later, our new dog, Freya, who would be asleep at my feet all these years later.

In both trees and books, it is said, small things lead to big things. What starts with the painfully slow slip of molecules across membranes – letters into words, cells into tissues – can reimagine a home, a life, the world. If re-storying our relationship with plants is as key to surviving the Anthropocene as I believe it to be, then drawing botany home might well be the most important work I will ever do as a botanist and artist committed to the plants of place. Come with me.

— PART I —

Roots
and
Shoots

POLLY'S COVE

BOTANY POND
DESOLATION SOUND
ADAMS RIVER
KAMLOOPS
ARMSTRONG
THREE-MILE POINT
LOPEZ ISLAND
LINCOLN COUNTY
MISSOURI RIVER
HELENA
FRASER RIVER
COLUMBIA RIVER
COLUMBIA RIVER
YELLOWSTONE
Pacific Ocean

Location Map for Roots and Shoots

1. The Web Below

CORRALLED BEHIND HORIZONTAL LOGS, the building hunkers in front of me, low-slung and panelled green. Neon beer signs blink in its high windows and across its exterior, white-painted letters spell out a name, JERRY'S SALOON AND STEAKHOUSE. Sitting in my bright-red Subaru, engine ticking toward cool, I am gravid and unsettled: weighted by the bulk of today, disoriented by the memories of the nights I once spent inside this building.

I probably shouldn't be here. Seven months pregnant, I should be at my desk finishing as much of my PhD thesis as possible before the baby comes. But my husband Marc needed to survey wetlands in this part of Montana, and I couldn't resist returning to the place I called home as a teenager. We spent most of the morning perched in the open horizon of a *Sphagnum*-dominated wetland called Little White Creek Fen; him counting plants, me drawing in my field journal.

Marc's wetlands might be part of what the map calls Lincoln County, but it isn't the Lincoln County of my memories. That one wasn't filled with plant names and the delicate outline of a grass-of-Parnassus flower. For the last two days, surprise at this landscape's vibrancy – the monkey flower and rein orchids rising above thick mats of moss, the dense spruce forest lining the banks, the red-tail hawks flying overhead and the sleek profile of a beaver coursing away from its lodge – has filled my field journal. It's not that this land isn't hard-logged. It is. Yet, between the clear-cuts, there's also an unexpected resilience.

But then, after lunch, climbing up through a narrow stream bed, I slipped and tore open the soft crease of my elbow. Yet one more scar from Lincoln County for my body to carry. As blood ran down my arm, I knew I needed a break from being pregnant in the field. Leaving Marc and our dogs for the afternoon, I drove off in search of a cool drink and air conditioning.

notes from the "little white creek fen"

8-21-2002

e v e n i n g   m u s i n g s

after this morning's short stop along paul crick fen, we spend much of the day just off Fortine-Wolf Creek road. we visit the fen lying alongside White Creek. Stunted spruce dot the carex utriculata. The parnassia immediately catches my eye, but I have trouble finding a dry-enough place to sit. finally perching on a hummock I am able to catch enough of the flower to satisfy me. when I try for a larger scale picture. last night's sleepless night catches up with me. I am joined later by Perra and Shasta. when marc returns, he finds all three of us napping in the light of an overcast morning.

The day turns strange when the subaru begins to make a terrible screeching noise from the right front tire. instead of heading directly to "little white creek fen" we descend further into the map of my childhood by driving into Trego to call for a tow truck. I am both startled and overwhelmed by my intense reaction to arriving into memories territory with anything less than complete control. It is with great, cloud lifting relief that I greet our return back into the land of wetlands.

now, with dinner over and the tunnel ventilation noise blessedly off - the puppies lay in exhausted heaps. a pileated woodpecker calls, white creek flows on in the distance and I feel so much more myself, caught within the context of a life I have spent the last 18 years building. I have come as a stranger to a land I once knew intimately

bryos from the fen

aulocomnium palustre
plagiomnium rugicum

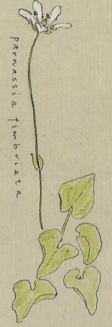

parnassia fimbriata

- my one drawing from white creek fen - the rest the morning I spent napping lying there on a dry hummo

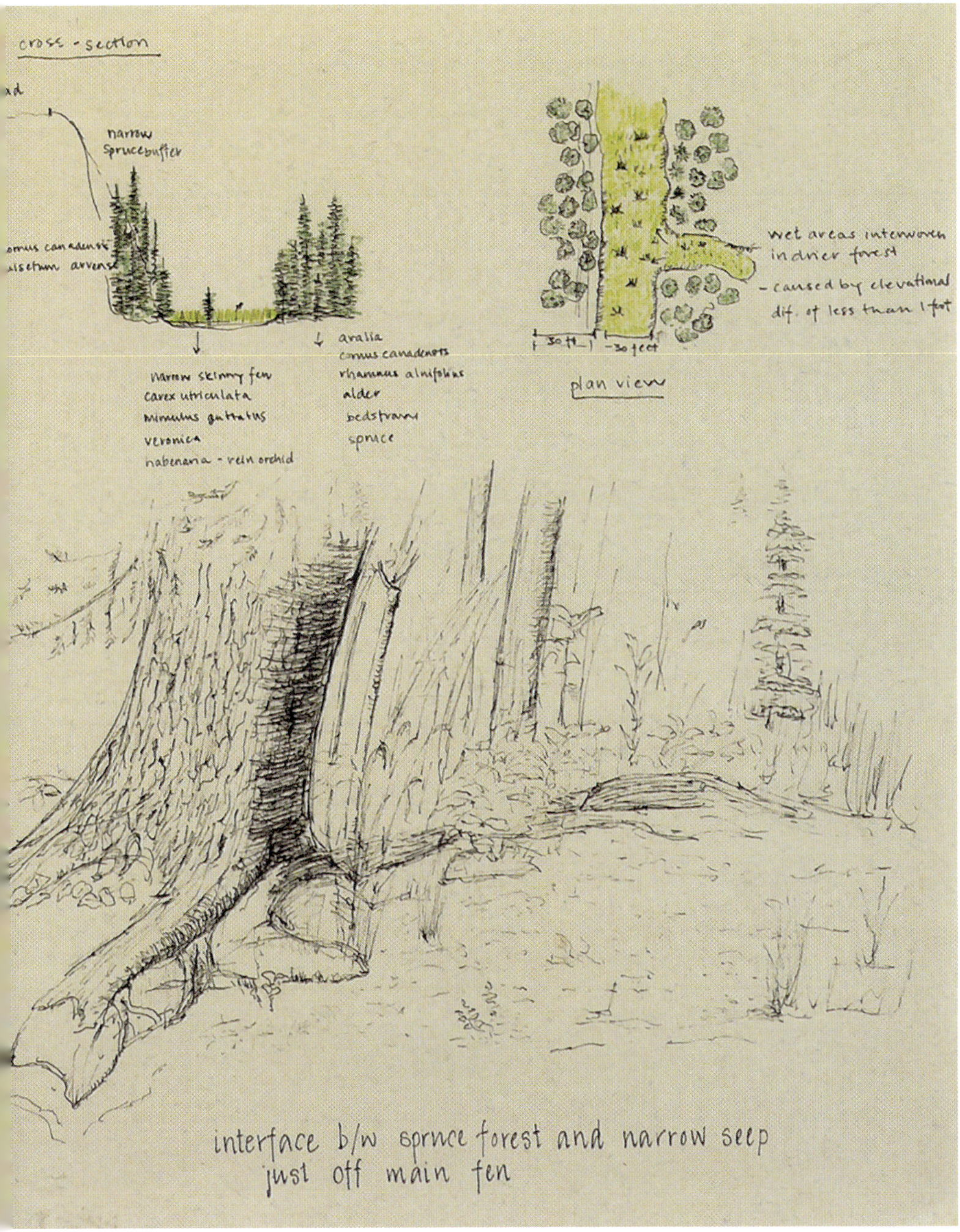

cross - section

narrow
Sprucebuffer

ad

omus canadensis
uisetum arvense

↓ aralia
    cornus canadensis
narrow skinny fen   rhamnus alnifolius
carex utriculata   alder
mimulus guttatus   bedstraw
veronica   spruce
habenaria - rein orchid

wet areas interwoven
in drier forest

- caused by elevational
  dif. of less than 1 foot

|30 ft.| |30 feet|

plan view

interface b/w spruce forest and narrow seep
just off main fen

Jerry's Saloon and Steakhouse: a small-town bar just off Highway 93, tucked into one end of a two-street cluster of homes and storefronts. Today, few people in my life understand the nearly mythic status of this place – both the building and its people – to me. Few people realize how my mother's new job bartending at Jerry's fed my family when Charlie McInturf left. Few people know I once carried its door key in my pocket, and that paycheques from its owner allowed me to move out of my mother's house while still in high school. Good or bad, known or not, there's no denying this building's space in my memories. Abruptly, I heave myself out of the Subaru's seat. Just this last week, the baby's bulk seems to have doubled, and every step I take feels like some version of waddling.

At the door, I hesitate. Out here, I feel secure, if rotund. Out here, I'm happily married; Marc and I have just finished remodelling our Craftsman bungalow in Helena, Montana's state capital. I have emails waiting for me on my computer and a schedule of tasks to complete. Out here, I'm a PhD candidate in plant ecology. In there...I don't know.

Maybe I should just go find Marc. But it's been over 15 years, and I can't resist memory's pull. I open the door, waddle across the threshold and into the narrow rectangle of light that precedes me.

Much is familiar. The cash register is where I expect it. The bar makes the same long slide to the right as I step past the liquor cage. But now, instead of one open space, the dining room on the left is divided into narrow booths. Slot machines stand in place of the jukebox. As I clamber up on the barstool nearest the cash register, I feel the slight slant of the bar underneath my elbows. I remember that – how the weight of elbows, year after year, had tipped the bar customer-side, and how you had to be careful setting down a drink that it didn't slide into the lap of a customer.

The woman behind the bar isn't unfriendly, but she's not anyone I know. When she asks, I tell her I'll take a 7Up, my order sounding out of place and juvenile, even with my obvious pregnancy. As she turns away from me, I glance right, along the bar's length, and am surprised by my disappointment when I see only strangers.

I turn back to the bartender, wanting to ask her if Dwayne or Kelly or Roger is still around, but she is already headed toward the other end of the bar. They must still stock the pop in the far cooler. And then the sight of her bending over to reach into the cooler releases a floodgate, and I am washed deep into the visceral memory of what it was to work this bar.

How walking the mat-padded runway between cooler and bar meant learning to rise with the onslaught of millworkers streaming in through the door; them tired and thirsty from eight hours on the second shift, you feeling necessity transform multiple gestures into one until you could use one arm to grab a Coors longneck out of the far cooler, the other to reach for the bottle of well whiskey, and your opposite foot to shut the cooler door behind you. How your body pitched and pivoted, glided and stretched, choreographed by need and repetition into an odd form of ballet. How the weight of verbal demands piled upon you. Three Rainiers and an order of those godawful-expensive fried mushrooms. A rum and Coke to go for the old lady at home, and two Buds for here. Lyn, can we get change for the pool table? Willie Nelson's "On the Dock of the Bay" filling the cement-floored room with sound. Learning to pour a tidy bowl – a shot of Kahlúa, overtopped with 151 rum and lit into flame as it was served. Adding numbers in your head, sliding bottles and glasses and cans across the bar, collecting money, and then three more orders en route to the cash register. Step and glide. Dip and bow. You never did go to your high school prom, but you got pretty good at this dance behind the bar.

And, for a dangerous while, you even revelled in it. You liked how the energy would build through the night, sometimes exploding in chaos like the evening Dwayne – filled with despair about some goddamn thing or another – threw his tidy bowl across the room and broke the jaw of the young guy from the seismic crew who had just stood up from his shot at the pool table. But more often whoever was behind the bar could lead the night's crescendo to the safety of last call. Some might say you learned this skill too early. That your mother shouldn't have suggested your name when she finally convinced Jerry to hire part-time help for her and the other bartenders on busy weekends. That a bar was no place for a 16-year-old girl, even if at the beginning it was only waitressing the steakhouse side.

But she did. And you jumped at the chance to earn regular money. And then, when you'd been working at Jerry's for more than a year, when the flood of millworkers ran too deep, your mom started you behind the bar, first just having you stock beer when the coolers emptied, and then teaching you to mix drinks. It wasn't legal, but the sheriff's deputies only came if there was trouble. And then the summer after your first year of university, your mom gave you her shifts and you walked the runway between cooler and bar alone.

You remember the first time you saw a line of coke, chopped fine and white. A group of local guys, home from the oil rigs in North Dakota, had spent all night in the back room, spending money like they were spilling water. They liked you and left a good tip – a fresh $20, maybe even a $50? You can't remember – but, as they left, one of them leaned over and whispered, "The rest of your tip's on the back of the toilet in the bathroom."

There it was: a white worm of trouble.

At first, you walked the mile home each night when the kitchen closed. But after you moved out of your mother's house, and lived with your best friend's family, you used part of your college savings to buy your own car, and you drove yourself. A 1964 Chevy Bel Air. Two years older than you; forest green with a white roof. You loved that car. Its dark green steering wheel and the shifter on the column. The broad, flat seats, upholstery worn but still intact.

The last year of high school was the worst. Especially when your university applications were in, and you were killing time, terrified they wouldn't take you, terrified that your only future lay in bars like Jerry's. Rage. That's the only word for it. You raged through that spring – waking up mornings without memory of the night before. The only thing you didn't bugger up was school. You made sure your homework was done before your first weekend shift started. Friday night, you still waitressed for your mom – the two of you working together in an uneasy détente negotiated out of financial need. And you remember that afternoon in April, when you came early to the bar with an acceptance letter from Bennington College in Vermont. You waited, skin itching with impatience, until Art Weydemeyer finally walked through the door.

It was your mom who'd introduced you to Art – tall and thin with thick, wavy hair and a face you thought looked vaguely European. It wasn't, though. Art was a Montana boy, born and raised. The only unmarried son of a big family, he'd come home from graduate school in California to run the ranch when his dad got sick. By the time you met him, his dad had already passed, but he still lived with his mom in the family ranch house. Art would come in early on Friday nights before the real drinking started. He'd order a Coors, sit at the end of the bar nearest the cash register. And if the dishes were caught up, and no one was ordering food, you'd lean on the end of the bar, listening to your mom and Art talk.

You grew to like Art. His family had been ranching in Lincoln County for nearly a century, and he knew the families of the kids in high school. You wondered how well he knew some of the mothers, but you never asked, and he never said. Sometimes he'd tell stories, but more often he'd want to know about you. Twenty years earlier, he'd taken the same high school English course you were in, and he'd ask, "Have you started reading Macbeth yet?"

And then he started bringing books for you to read. There was one by Jerzy Kosinski that hurt to read, and that one by Joan Didion, and that story about the guy turning into a cockroach you're pretty sure you never finished. But there was also that book by Loren Eiseley that made you feel as though you could be part of something bigger.

And then, at the beginning of your senior year, when you were trying to figure out which universities to apply to, Art suggested adding Bennington College to your list. He'd read about the college, he said, in one of his magazines, and it sounded interesting. You remember rehearsing your ideas with Art across the slick surface of the bar before writing an essay comparing Eiseley and Kosinski for at least one of your college applications. In late winter, when Bryn Mawr College asked you to interview with their recruiter in Missoula, Art drove you the four hours south and stayed by your side during the introductory tea.

Of all the people in your life, Art was the only one who knew the different paths you walked – your nights at the bar, the sweaty hallways of your high school, and maybe even the life you were hoping to find. Art also knew the grim reality of your finances; how much you needed not just an acceptance letter, but a substantial financial aid package.

The summer you came home from Bennington, you'd open the bar at 4:00 p.m. and sit on the logs outside, waiting for the first customer. More often than not, it was Art – fresh from the hayfield, gloves still tucked in his back pocket. You remember the day he asked about your Bennington classmates, and you tried to describe how different they were from you. In your poetry seminar, you said, there was even a girl named "Quintana."

Sitting beside you on the log, drinking his Coors, Art had choked in surprise. "Quintana?"

When you said "Yeah, Quintana Dunne," Art had set his beer on the ground, told you he'd be back, walked to his truck and drove off. Fifteen minutes later, he returned, carrying a small book.

Titled Quintana and Friends, *it featured a small girl in a floppy hat sitting cross-legged on the front cover. Art wanted to know if you recognized the girl.*

"Maybe." You weren't sure.

"Well, if that's the same Quintana, then you took poetry with Joan Didion's daughter."

And then, after you went back to Bennington at the end of the summer, your mom moved with her new boyfriend to Washington state and, for many years, you had little reason to return to Lincoln County.

THE BARTENDER is back in front of me, asking if I want anything else. No, I tell her.

Sitting on a barstool on the customer side of the bar, with a baby turning somersaults inside me, it feels impossible that no one here knows this place was once part of my home, that no one in this building knows what it took for me to leave.

I stand up, pull my wallet from my field journal bag, and lay some bills on the bar. I need to make a phone call. Outside in the hot sunlight, I squint and put on my sunglasses. I have just enough time to call Art before I need to go pick up Marc and the dogs.

WHEN MARC DROPS ME OFF the next morning, Art's waiting for me outside his ranch house. The inside – an interior space that occupies no place in my memory – is shaded and cool. Art offers me a tall glass of lemonade. As we sit down, I notice the books and magazines stacked by his chair. I'm nervous, worried that after so many years Art and I will have nothing to talk about. I tell him how surprised I was to learn, shortly after Marc and I moved to Helena, that in between raising cattle, his family had helped protect Montana's wild areas. That one conservation group is trying to establish a mountainous region just north of here as a national wilderness in honour of his uncle.

And Art – older, but the same self-effacing man I remember – says, "Well, you never asked." He's glad I stopped working as a natural history guide on cruise ships. He doesn't say why, but I understand he thinks guiding carries some of the same risks that bartending did.

When I ask him if he still goes to Jerry's Saloon, he says, "No. Never."

Talking to him, I wish I'd kept reading literature, just so I could have more to share, but he seems to enjoy my stories about teaching undergraduates. I tell Art that I think teaching uses some of the skills I learned behind the bar at Jerry's. These days, I spend more time in classrooms than in saloons, but I've yet to meet a professor who can read people better than a bartender.

Art tells me about his sisters and brothers, and how they still come home each summer to help with the haying. I have only dim memories of his siblings, and our conversation is muted. I realize this is the first time I've ever spoken with Art inside his house. It's been a long time since I left for Bennington. By the time Marc drives up the driveway, I'm glad I've come, but I know it might not be easy to visit again. As Marc and I drive south along Highway 93, we speed past miles and miles of clear-cut blocks – some are newly shaved, others are straggling up into pole-sized stems. It's still hard to be a tree in this part of Montana.

Less than six weeks after Marc and I return to Helena, our daughter Maggie arrives in a flood of pain and emotion, and she is perfect. Perfect.

TEN YEARS LATER, I stumble across Art's obituary. I read that when he came home from graduate school in California, he juggled his thesis with ranch work, and that he loved to fly-fish, and he was renowned for the tameness of his cattle. At the university where I now teach, my botany students and I are reading a paper about the mycorrhizal web found within the forests of the Pacific Northwest. No tree lives without risk, but within these forests, this interlocked web allows those who came before to care for those who come after.[1] Sugars, nutrients, water, even medicines, flow from one root to the next, crossing from old to young, from well-nourished to needy, from one species to another.

I met Art when I needed to leave Montana. Within the smoky, dim-lit interior of Jerry's Saloon, Art taught me to value well-told stories, to believe in the rhythm of poetry, to wonder at the spare elegance of a Joan Didion essay. But, outside Jerry's, Art and I inhabited very different Montanas. Whether he wanted it or not, his Montana centred on a family ranch, replete with history, that tied him to the land. His Montana knew to alternate the obligation of orphaned calves and hayfields with fly-fishing trips in the mountains. My Montana depended upon the precarious finances of a single mom. My Montana was poor in tradition, rich with men made dangerous with beer or Black Velvet. In my Montana, leaving was the smart thing to do.

Today, I regret that I never asked to go fly-fishing with Art. I think I would have learned as much from him in the mountains as I did inside Jerry's. Here's the hard thing: I know what gifts Art gave me; I'm not sure I ever did a good job returning his kindness. But maybe it's not about reciprocity. Within any fungal network, nutrients flow along gradients of need, not paths of indebtedness. Tall, well-nourished trees transfer more sugar to shaded seedlings than to those in the sun. Overall, the resilience of any forest lies as much in its ability to care for those in need as it does in the size or growth of individual trees.

Art's obituary says that a university scholarship was established in his name for "students living in northern Lincoln County interested in outdoors and ecology." I don't know if I was the first ecology student supported by Art's generous nature, but I take some comfort knowing I won't be the last. In my botany lecture the next morning, I look out across the faces turned toward me and think about the threads of learning it is my job to support. I think about the many ways stories are shared – across the slick surface of a bar, wet-footed in a *Sphagnum* fen, within ordered confines of a lecture hall – and how good stories, if told well, outlast the storyteller.

What I know for sure, as I begin the slip and slide, the pitch and pivot, of a brand-new lecture, is that all of us are supported by those not related by blood. All that we can ever do is to keep contributing, unasked for and without acknowledgement, into the network that binds us all.

2. Nutrients from Away

IN THE MISTY LIGHT, my van door shuts with a thud. I stand on a gravel parking lot hanging above the Adams River, in a space empty but for my van and me. It's Tuesday, early morning, late October. I should be on my way to my office, but I've rescheduled meetings and postponed deadlines to slip out of the house before Marc and Maggie were awake, and drive 45 minutes northeast, out of the sagebrush steppe of the South Thompson into the wet interior forest of the Shuswap Watershed. But now indecision stalls my arrival. Up or down? From the parking lot, one well-signed trail leads to the Adams River. Across the road, another trail, less obvious and technically closed until BC Parks can fix its collapsing bridges, leads upstream along Hiuihill Creek.

Toward the margin or the main current? A large pickup truck, and then a minivan – folks on their way to work – splash around the corner, breaking my indecision. I walk across the road, slip down the narrow track and head upstream. This is my third trip out to the Adams River in as many weeks. Beneath my feet, birch leaves moulder into compost, and above me the sky is lumpy with cloud, drizzly with rain. In the silence, the landscape drips into place around me. This corner of BC – bisected by river and lake, caught within the clouds pushed upward against the Monashee Mountains – may be less than an hour from my home in Kamloops, but its ecology is several zones wetter.

Rich in cedar and Douglas fir, birch and cottonwood, with a thick undermat of *Polytrichum* and *Hylocomium* mosses, the botany of this forest is always luxuriant. But, every four years, the return of its sockeye salmon swims this landscape into something more. Two weeks ago, on Thanksgiving weekend, hordes of people attended the Salute to the Salmon festival on the broad delta of the Adams River. Beside a parking lot overflowing with cars and buses, vendors sold bannock, hotdogs, kettle corn, mini-donuts and artisan crafts. Nearby, local musicians and First Nations storytellers performed on stage, while posters in the Ministry tent detailed the salmon life cycle. Two weeks ago, this landscape swarmed with bodies gathered from around the world. Now it feels nearly abandoned.

On the trail headed upstream, I think about how the confluence of choice and chance shapes our willingness to swim *with* or *against* the current. All month, I've been compelled by the aquatic drama unfolding in the Adams River, but I've little to explain my obsession. I'm neither an aquatic biologist nor a weekend fisher; I've not thought much about fish in more than two decades. But something has pulled me back out to the Adams today – to walk away from the main run by myself, days after a month-long festival attended by nearly half a million tourists.

Conformity, contrariety. Biology swims with both. Certainly, little smacks more of conformity than the red bodies filling this stream bed. Millions of sockeye swim the same direction, pulled by the same urges, destined for the same fate. For most, each stage of life – eggs, fry, smolt, returning adult – has been swum in the close company of others. Beginning as fertilized eggs in gravel four years ago, the majority spent one year growing in Shuswap Lake before migrating out into the Pacific. Further conformity ensued when more than 90 per cent took the same interval – four years – to sexually mature and return home together.[1] In the Fraser Watershed, the sockeyes' conformity with one another means that hundreds of thousands, sometimes millions, will return together as a dominant run once every four years. In the intervening years, their numbers thin to thousands, or even hundreds. A pulse with four beats: one strong, three tiny. All, today, facing increasing risk.

At the next fork, I take the smaller path, ducking under low-hanging cedars and walking through a seepy area to the creek's edge, where the water gathers into a shallow pool. Two weeks ago, this pool ran quiet; today, it splashes red. I count three, no four, five bodies in the water. Bodies that travelled nearly 400 kilometres against the current of three rivers – the Fraser, South Thompson and Adams – and one creek, Hiuihill, to reach this spawning pool. When I walked the main stem of the Adams River earlier this month, the sheer abundance of bodies made it difficult to focus on separate fish. But here, at the limits of their spawning territory, the run resolves into individuals. Settling down on the bank, I shrug off my backpack and pull out my sit pad and field journal.

In the pool, the bodies of both males and females are scarlet with carotenoid pigments, but the backs of males are distorted into humps, their noses contorted into black-edge hooks above sharp teeth. Two males, appropriately called hooknoses, hover near females, while the third male lingers alone – a satellite with neither moon nor planet to orbit. In the pool, long moments pass where the only visible movement is the idle ripple of pectoral or caudal fin. But then one female rolls to her side, tail thrashing a trough in the gravel below, and the satellite male rushes forward, only to slide back downstream when the female's dominant male pushes him back. Within seconds, the three fish have resumed their original position and stasis returns. In the silence, I scan the bank, the downed logs, looking for more.

Conformity can be a matter of degree. In this river, dominant and satellite males use variations on a theme – large size, physical force – to squirt their milt atop newly laid eggs. Far more rare are those contrary, undersized-males called *jacks*. Typically, these males – one-half the size and a year younger than most dominant males – rely not on force but subterfuge and opportunism to pass on their genes. Most of the time they lurk behind rocks or logs, sneaking out to squirt milt only when fighting distracts the larger males. I've read about these males, and last week, I saw a few jacks spread out on a white table by fisheries biologists, but I've never actually seen them in the river.

ALTERNATIVE STRATEGIES fill the river of biology; as a hippie kid in southern BC and northwestern Montana, my position outside the majority was beyond my control. I was raised on a contemptuous view of Nixon's silent majority. As a teenager, at work in Jerry's Saloon, I was the waitress struggling to read Kafka. My acceptance by Bennington College permanently cemented my outsider status. Few enough of us who graduated from Lincoln County High School in 1984 wanted to go to university; no one else I knew had their heart set on an odd liberal arts college on the other side of the continent.

In the months between my acceptance and arrival, did I envision Bennington as the stream in which I would swim with the current? I'd never set foot on Bennington's campus, but I knew the basics: renowned for its creativity, Bennington institutionalized rebellion over acceptance, *doing* over *memorizing*. I've no memory of deciding between the University of Montana's smaller but more than sufficient scholarship package and Bennington's much larger but insufficient package. What

FIELD OBSERVATIONS:

purple margins of Douglas-fir bark furrows. Disporum leaves folding into flesh colour. Mahonia. hazelnut. chickaree. nuthatch. mock-orange. spiraea betulifolia. flicker calls. Lonicera ciliosus. rubus parviflorus. Rhamnus purshiana. Elderberry. thimbleberry. Eagles 'ki-ki-ki' in the distance. cottonwoods gleam. rosehips burn with anthocyanins. juvenile bald eagle soars. train sounds cross the valley. fish hang belly up while others — red bodied and green headed — do what they need to do. Side channels embrace spawning fish and in their surface marine anthocyanins splashes with terrestral chlorophyll & carotenoids

red-osier dogwood

paper birch

Leaf colour transforms and then falls: Cascara,

high bush cranberry

cloudy, no rain, stiff wind from the east •   o c t o b e r   25 • 2

The memory of lignin at Tsútswecw Park carries the imprint of human hands crafting as the nutrient flux carried by salmon.

In the side channels, the red fish become nearly terrestrial, writhing their bodies across exposed rocks. Each time, I watch their acrobatics, my heart nearly stops. So much hangs in the vertical balance, in the muscle and tendons, of these bodies poised between life + death, between sea + forest. So much depends on the tenacity, the will, of those who have not eaten in weeks, who come home to reproduce, who come home to do.

This year's drought has left some channels bone dry. Sitting in the squeak + squawk of this ecosystem, I trace this year's against the four previous dominant runs I can remember. Red, yellow, rotting flesh, white gulls clustering. All magica but so muted against the memories embedded in the lignin of this river's forest.

2006 ∘ 2010 ∘ 2014 ∘ 2018 ∘ 2022

I think about now, more than 30 years later, was that, if I did consciously choose, I didn't know what I was choosing between.

Certainly, nothing prepared me for the eastern deciduous forests of our continent. In late August, after stepping off an Amtrak train into Troy, New York, where tall, skinny houses butted up against narrow sidewalks, and then riding in a shuttle van into the wooded hills of southwestern Vermont, I struggled to follow even the simplest conventions of local geography. The hills surrounding the college, I was told, were the Green Mountains.

Mountains? No one I knew would call such restrained topography "mountains."

Green? One monosyllabic adjective felt inadequate to describe the layers of photosynthetic pigment that climbed through tree canopies, spread across intervening fields and clawed at the roadsides. Even the air was wrong. A humid thickness, composted with leaf mould, rather than a crisp dryness, accented with dust and pine pitch.

Chance and choice. Arriving at Bennington College, I occupied a minority position as the scholarship kid from the intermontane west, but why did I choose to study science at a school renowned for its visual and performing arts? It wasn't something I'd intended – I'd only enrolled in first-year biology after a poetry class was cancelled. But I do remember when it happened. Outside the Dickinson Science Building, the leaves on the maple tree were falling crimson red. Inside, Betsy Sherman, a vibrant, curly-haired woman who showed slides of her blonde, chubby-cheeked sons as evidence of her own evolutionary fitness, was describing the ability of viruses to hijack the machinery of human cells. I was following, but only just.

Viruses used our DNA to replicate themselves?

"Holy shit," I exploded, indignant and amazed. As the entire class turned to stare at me, Betsy burst into delighted laughter. But not even the flame of embarrassment that ran up my cheeks could quell my intrigue.

In biology, decisions get made. Some are hard-wired, controlled by genes. Many others are contingent, dependent upon the waters in which we swim. The sexual maturation of jack salmon contains both. Early puberty depends upon specific genes being turned on; a triggering that only occurs for those males that find food enough to grow large at an early age. Hard-wired or contingent, all decisions have consequences. If the size of jacks limits their sexual strategy to sneaking, it also minimizes their risk of being eaten.[2] Not only do these salmon escape the ocean's predators early, but their small size makes them less attractive to the bears who patrol their spawning streams.

For any organism, each new pool swims with both opportunity and constraint. At Bennington, my science was immersive but small. I worked with Betsy, the animal biologist, teaching grade school children to design their own experiments; gardened with Kerry, the plant ecologist; tested native plants for their antibacterial properties with Tom, the organic chemist. Today, there are famous names I ate alongside in Bennington's dining hall – students like Donna Tartt, Bret Easton Ellis, Peter Dinklage, even faculty members like Jamaica Kincaid and Edward Hoagland – but I have few memories of those whose orbits never brought them into the science building. The year I graduated, only eight of us stood up from natural science and mathematics.

In front of me, five salmon punctuate stasis with frenzy – again and again. I lose track of the number of times the satellite male attempts the salmonid equivalent of a shoulder tap, only to slide downstream with a dominant fish in a male-only promenade. Once or twice I think I see a jack lurking under the cutbank, but each time I mark it up to wishful thinking. Later, when all the final counts are in, one of my neighbours who helped monitor this run will tell me that only 40 jacks were counted amid the nearly one million sockeye salmon that returned.

If I was unprepared for the East Coast's deciduous forest, I was even less prepared for my classmates. That first morning on campus, in line for my student ID, I stood behind a gregarious, lean, dark-haired guy I'd first noticed on the train from Chicago. He would be in my calculus lecture, and would later finish Bennington to work as a professional dancer and tree arborist. Eventually, I would learn his name – Jack – but in the moment I couldn't stop staring at the young woman processing our forms. In between asking for names and pointing out where to stand to have your picture taken, a strange form of dance was occurring between her and a chair. She was, I imagined, working on a choreography. All this, or something similar, I think I expected. But what I did not expect – what was unimaginable to me – was the startling expanse of this young woman's cleanly shaved head.

Today, out beneath the birch and cedar, beside the running waters of Hiuihill Creek, my remembered reaction to a young woman's fashion sense makes me smile. I fancied myself such a rebel; it took only one young woman's smooth and oiled skull to reveal my cultural biases. When I chose to study science at Bennington,

I migrated away from its majority, but I still swam within its current. Science at Bennington had few requirements other than mandating that its students study beyond its boundaries. In between studying immunology and physiology, biogeography and chemistry, I learned to draw architectural plans, stumbled through French lessons, practised the steps of African dance, and read poetry. Bennington was a river that, by design, expected its students to dance or sing or calculate *across* borders.

Suddenly, I tire of my solitary existence on the margins of this run. Gathering my gear, I walk back through the forest to my van and drive the few kilometres down to where the Adams River spills into Shuswap Lake. Gone are the cars lined up to get into the park; in the parking lot built for thousands, less than a dozen cars and one school bus squat in the corner nearest the river. Two men clad in bright orange rain slickers use drills and pry bars to deconstruct the plywood shelters that earlier housed crafts and T-shirts.

I walk down into the thick of the salmon run, to where colour lives not just as reflected light, but as a smell and a weight emanating from the carcasses littering the stream edge. Out on the black metal platform that protrudes over the Adams, I am alone. In the forest canopy, cadmium cottonwood leaves dance alongside olive-green cedar. Bald eagles soar in distant sky; gulls float on river water, heads probing below the surface for exposed salmon eggs. Two photographers kneel on gravel in the far distance, and a kingfisher swoops overhead, rattling as it goes.

Beneath me in the river, the decaying salmon are no less marvellous than the viruses I first studied at Bennington. The odds that any one of them are here are simply astonishing. For every 3,000 eggs fertilized and covered in gravel, at most one spawning adult will return. One life, that to be successful, will have to physiologically transform itself. Twice. Outbound and inbound, in that brackish boundary between fresh water and salt water. In each direction, salmon pause for several days, even weeks, and in their hesitation, their kidneys develop new setpoints for urine concentration; the cells in their gills reverse the direction of their salt pumps.[3]

At the level of individuals, this transformation is the physiological rejiggering that allows salmon to benefit from both sides of the freshwater–saltwater boundary. Eggs and embryo develop within the protected environment of river gravels. Maturing fish use the rich resources of the ocean to grow rapidly. But, at the level of ecosystems, the journey encompassed by this transformation does nothing less than stitch together land and sea.[4] Beneath my feet, the bodies of the decaying fish are muscled with marine protein, coloured with carotenoids filtered from marine plankton. For the last month, eagle and bear, wolf and marten, have been pulling salmon from the river, eating what they want and letting the rest rot beneath cedar and Douglas fir, birch and cottonwood.

Looking out along the trees that crowd the mouth of the Adams River, the understory of hazelnut and thimbleberry, the ground layer of *Polytrichum* moss and bunchberry, I wonder how long it will take for the nitrogen ferried via salmon bodies to slip upward in cedar and cottonwood. One year, two, five? Do the trees, I wonder, worry about the chance and choice that brings a salmon carcass near its roots? Now, when this year's run, against all expectation, has returned, and the bear and the marten and the eagle feed, do the trees celebrate the rich pulse of nitrogen that will flow into the web holding this ecosystem together? From carcass to maggot to bird to song? From soil to root to seed? Even now, with the abundance of this year's decaying flesh, do the trees worry about what's to come? Do they ache for years ahead, when river waters will grow too warm for the cold-water fish that have fed them since time immemorial?

Out on the platform, the rain begins again. It's time to leave, but I want one last circuit on the trail that leads out to Shuswap Lake. I think about the tiny sockeye fry that will emerge from river gravel come early spring. Of those that return to this forest, most will be big; only the rare will be jacks. At the level of individuals, it's nearly impossible not to focus on their differences. But walking within a forest fed by salmon, I can't help but marvel at the song that gets sung across the differences. There's a reason the first stewards of this land, the Secwépemc People, named two of their months for the arrival and then the spawning of these salmon.

No botany, I realize, grows without nutrients blown, washed or swum in from afar. The decisions we make – to haunt the margins of a salmon run, to study science at a liberal arts college – always reflect both *who* and *where* we are. Choice and chance – both are part of our experience of the world; both are woven into our individual biographies, our propensity for one strategy, one position, or another. All life is chemistry shaped by experience, but in experiencing multiple worlds, in crossing boundaries, these salmon help sing my world and its botany into being. How long, I wonder, will any of us last if their part of the song is no longer heard?

On the trail, the sound of moving water grows louder as I near a turbulent drop-off. Just beyond my feet, water floods past red fish manoeuvring against the current. Once again, there are too many to track individuals, but then one fish turns belly-side up and is swept away. For just a moment, I let myself imagine the force of the water, its sharp coldness against already numbed skin, the weight of fungi spreading across cells, the final dissolution of flesh. Big fish, little fish. In returning, one fertilizes nitrogen-limited forests, while the other spreads genes across populations segregated by year. Big fish, little fish; in returning home, both transform themselves and their ecosystem.

3. Seabound

"EVERY BOAT CLAIMS YOU."

That's what I think, standing thigh-deep in the salty water of BC's Desolation Sound. In front of me, the sea mirrors the day. On its surface, *Fucus*-bronze, cedar-green and *Racomitrium*-gold dance alongside sky-blue. In the kayak beside me, my dry bags are packed, food measured into daily allotments, clothes and sleeping bag double-wrapped, camera and field journal locked in a waterproof Pelican case. It's time to go. Marc and impetuous, 11-year-old Maggie are already paddling away. I wonder if they paused moments ago as they clambered into their double kayak, or if hesitation lies just within me, an indelible stain of the ten years I spent going to sea.

I started out a reluctant sailor. Less lured than press-ganged by the scarcity of jobs for novice biologists, I boarded my first boat in Dutch Harbor, Alaska, the main port for the Bering Sea fishery, on a gray January day in 1990. Climbing the metal gangplank, I dragged behind me the bright baby-blue sampling totes that marked all of us who went to sea not as crew but as the government-mandated guests known as fisheries observers. My decade at sea wasn't continuous, nor was it on the same vessel, but at least once, sometimes as many as four times a year, I would step from stable ground to shifting platform, from dock to deck. Some voyages lasted three months, others were as short as a week.

However we go, I think, as I push off into deeper water, balancing my weight above my kayak before sliding legs into cockpit, bum into seat, it's the boat that's the thing.

Today, we're not going far – only three days of paddling – and Marc and I have been here before. But given our terrestrial biology, hesitation seems reasonable.

Going to sea. It's an odd phrase. It implies that the sea is a destination, but you can't go to sea in the same way you *go to Jack's house.* When our ancestors dragged themselves from the water millennia ago, they started on a path that would leave us addicted to atmospheric oxygen. For terrestrial animals, going to sea means perching atop temporary platforms of existence. Without the ability to filter oxygen from water, full immersion in the sea results in the suffocation we call drowning. We may carry sea water in our cells, but we can't have it in our lungs.

The irony is that, once ashore, terrestrial organisms never stayed in place. Chance, made physical in flood or hurricane, regularly pushes land animals back into the sea. Most die; a lucky few survive. In 1995, fishermen found 15 green iguanas clinging to the surface of hurricane-uprooted trees on the beach of Anguilla Island in the Caribbean. These terrestrial lizards must have spent at least three weeks drifting at sea before coming ashore 150 miles north of their known range.[1] Even plants, as seeds, can travel oceans. Nearly 10 per cent of Hawaii's vegetation is believed to have rafted to the islands on large mats.[2] And humans? By accident or with intent, going to sea is one of our oldest stories.

I NO LONGER REMEMBER how many different fishing boats I worked on in the Bering Sea. I do remember the first: a 280-foot factory trawler called the *Starbound.* A big boat, nearly brand new and housing a United Nations–like crew: a German engineer, an American captain, a Norwegian fish master, Japanese quality-control techs, deckhands from Seattle, factory workers, including a dozen or so women, from across the United States. This crew of more than a hundred was diverse enough to absorb my presence and my sampling gear – totes, metal scales and logbook – with few ripples.

I remember the best: the *Patricia Lee.* By the time I boarded this boat, I'd already spent nearly two years working as an observer. I'd learned to brace myself against the pitch and roll of a Bering Sea storm, and to linger in the light on deck whenever I could. I knew how to measure cod-ends jam-packed with pollock on icy trawl decks, the best way to sex and measure blue king crab in cargo holds, and how to finger tally the golden-specked bodies of Pacific cod that came lumbering up on longline hooks. I'd even survived a stretch on one of the worst draggers in the fleet – an enormous catcher-processor with a dirty crew that fished even dirtier.

My hesitation the March day I boarded the *Patricia Lee* was caused not by the boat, or by its all-male crew, but by *where* she was headed. To reach the *Patricia*

Lee's fishing grounds from Dutch Harbor, we would steam nearly 1000 kilometres west, trading today's western longitude for tomorrow's eastern designation. There, past the International Date Line – in a territory centred on the volcanic Semisopochnoi Island and an underwater feature referred to as "the Hump" – the boat would pull and process brown king crab for 90 days without returning to port, supplied only by itinerant tramp steamers that would deliver fuel, mail and groceries in exchange for crab frozen into product. From March to June: 90 days without touching land; without access to a phone; without saying hello to a stranger.

Paddling behind Marc and Maggie, out past the long, linear sandspit that envelopes this tiny bay, it's not our isolation but the shift in perspective that startles me. I'd forgotten this. At sea, your horizon is the boat's horizon. There are no hills to climb, other than the ones made of wave. Even on the coastal chart strapped to my deck, the low-lying ridges surrounding our launch site pale in comparison to the lines of relief extending beneath the water's surface.

At sea, we are beset by motion beyond our control. On fishing boats, I came to expect my body's temporary rebellion to this altered perspective, knowing I would spend the first 24 hours curled into a comma of misery on my bunk. Even when not seasick, your profile – that simplified drawing of self – becomes subsumed. In front of me, Marc and Maggie have become an elongate line, with one small and one big bump positioned amidships.

When I finally catch up to them in the main channel of Okeover Arm, we pull the boats side by side to look at the chart. Marc and I decide to cross from one side of Malaspina Inlet to another, intercepting a small islet, Lion Rock, on our way. I wonder if leaving the shore will make Maggie nervous, but before I can ask, Marc and Maggie paddle out of easy conversation range.

GOING TO SEA complicates communication. In the close quarters of a fishing boat, banter rises, slips and falls. Stories get retold, again and again. Late at night, when pots rise – empty of crab, full of bait – nearly any syllable risks offence. More often than not, the words of men dominate. During my time aboard the *Patricia Lee*, the only female voice I heard – other than my own – was on the radio. Peggy Dyson gave the weather report twice a day for the Bering Sea, area by area, over the single-side band radio from Kodiak, Alaska.

Hello, all mariners. This is WBH-Two-Nine Kodiak. Today is Wednesday, May 5, the time is 0800 hours, Alaska Daylight Savings Time. This is the National Weather Service Marine Forecast, valid from present time, 8 a.m. Wednesday to 8 a.m. Thursday.

Not only would Peggy stand by to repeat the forecast to the gaggle of voices that followed – "Area seven bravo, repeat please," "Area five, repeat please," – but her words were echoed, deep voice to deep voice, to those beyond the reach of her radio.

Other times, words carved space for me. When I first met Mike, the skipper of the *Patricia Lee*, I thought he was too beautiful to be a fisherman: James Dean eyes; long, thin body; worn Levis tucked into short leather boots. He'd come north to Alaska to make the money for a graduate degree in painting, only to end up as captain and part-owner of a crab boat. Mike had taped photos of his wife, Alexa, and new son, Luc, on the instrument console above the captain's seat. It was only near the end of my time on his boat that I learned Luc was his second son. His first son, Mathew, born to a previous girlfriend, had died in his crib on Kodiak Island. Every landscape – even isolated platforms of wood and metal and men – I realized, was awash in stories.

IN DESOLATION SOUND, we are well past Lion Rock when I realize a different story has caught up with me. Pulling into a small, rocky cove, Maggie and I leave Marc with the boats while we clamber up a talus slope of moss and lichen-covered rocks until we are beneath the cedar trees. Of course, it is only when my shorts are down – Maggie's too, I think – that a large, crashing sound comes from the ravine to our right.

I look up. Marc is staring at the ravine. The noise is big and headed downhill. Clearly, it will make it to the boats before Maggie and I can. I'm still trying to figure out what to do – other than pull up my shorts – when an undulating line of fur, complete with an impossibly long tail, comes into view. A river otter.

Close encounters with beings of another kind. No sea journey is without them; all of them immerse you in the immediate. On the *Patricia Lee*, I learned to cradle least auklets, trapped in the processing room, in my hands until the deck crew finished setting pots. It was the first bird species whose plumage I learned in hand, rather than through an illustration. Each time the boat moved from one string of pots to another, there was the chance of Dall's porpoises. Peering down from the shelter deck roof, I would watch their sleek bodies slip and slide in the bow wake, feeling encircled in their joy. Who could resist the sea otters, lying on their backs, sharing with their pups the crab bits tossed to them by the crew? Even the brown crab, with their long, spidery legs and ruby red eyes perched on stalks, pulled at my imagination. What was life for them, crawling on the ocean floor, fathoms of water beneath us?

Back in our kayaks in Desolation Sound, the river otter pops up in front of us twice more before we paddle out of its company. It's not far to tonight's camp at Hare Point. Marc and I have planned this trip to be short on paddling time, long on *puddling* time. All of us, but perhaps Maggie most of all, love tide pools. We're not disappointed: in the rocky expanse beneath our campsite, the shapes of urchins, sea stars, crabs, mussels, barnacles and spiny cucumbers challenge our understanding of life's possibilities. The sea has always been a cauldron of reimaginings. Even today, long after life has evolved from its earliest beginnings, there are many more ways to be an animal in the sea than on land.

After dinner, the incoming tide chases us onto the granite rocks knuckling up out of the sea, and mosquitoes drive Marc and Maggie into the tent. But I'm reluctant to abandon the long light of a summer evening. Just below my dangling feet, marbled murrelets raft between bobbing seals. Farther out, ospreys plunge into a placid surface.

Tonight's sea might be flat, but I don't trust it. Every sea – even a sheltered one – is a twitchy beast. When I look at the water closely, I see restless currents, driven by tide, streaming around Hare Point's rocky protuberance. I'm glad I'm on land; it wouldn't take much wind to prod this tidal rip into standing waves. Not often, but more than once, the turbulence of the Bering Sea sent the *Patricia Lee* scurrying for the lee of Semisopochnoi Island. Other days, the same water settled down, flat as a mirror. Mike called such days "floating through sea" days. When he showed me photographs of his paintings, I asked him if the Bering Sea ever crept into his paintings.

"Do you mean do I ever paint boats? No, I never paint boats. But the light, I always take the light home with me."

Here in Desolation Sound, gold races before dark, its retreat illuminating the island opposite me, Josephine Island, with an inner glow. I race to get it down, scribbling graphite, slapping pigment, on my field journal page. I stay as long as the light, filling the pages of my journal with sketches of seals and shadows, islands and arbutus trees, before climbing back up toward Marc and Maggie.

Close to our tent, I hear Marc reading aloud to Maggie. I settle on a rock, wanting a bit of the story before I interrupt.

Some people go to sea with mysteries or potboilers. My husband brings Homer. Maggie has caught her dad's enthusiasm, and has her own favourite parts of the *Odyssey*. Tonight, Marc is reading from well into the book: Poseidon's wrath has forced Odysseus from the limited security of his hand-hewn raft, and Odysseus is adrift in waves, struggling to reach the shores of Scheria. When Odysseus finally makes it through the breakers and falls asleep, safe in a "fine litter of dead leaves,"[3] I unzip the tent.

On land, I've never been a night owl. On the *Patricia Lee*, I learned to inhabit the dark, one or two hours at a time. The day I boarded, Mike asked if I would stand wheel watch along with the crew. The idea was terrifying, but I'd agreed. Each hour I spent behind the wheel would be an extra hour of sleep for a crew who rarely got more than five or six a night.

Wheel watch depends upon routine: wake, shaken out of sleep by calloused hand, stumble to the galley, pour coffee, slop milk, climb with cup in hand to the wheelhouse and clamber into a seat still warm from the man who just woke you. Drive the boat in circles, following the green line plotted by Mike before he went to sleep.

At the end of the hour, check instruments, descend into the roar of the engine room, check gauges, return to main deck, wake whoever's name lies below yours in the watch list taped to the instrument console. Climb back up to the wheelhouse. Wait.

Routine, yes; but something like magic hovered in those Bering Sea nights. Clouds passed over the moon. Underneath me, the boat swam through blackness, its movement mirrored in the green light of the instrument panel. Auklets zipped through deck lights, while the snores of men – Mike in the cabin just off the wheelhouse, the rest of the crew in the deck below – blew counterpoint to the soft sounds of Sinead O'Connor, John Prine, Bruce Cockburn, the Waterboys, Daniel Lanois.

It was in the dark – driving someone else's boat, alone but for a few lights in the far, far distance – that I began to understand how we are shaped by those we seek. Crab or auklet, moss or tree: all species inhabit their own worlds. Whenever we go in search of species other than our own, we entangle ourselves in the lives and places of others. Fishing brown crab meant abandoning tree and soil, home and family, to float for months in isolated precarity. Sometimes it was too much. Midway through my time on the *Patricia Lee*, a crew member refused to leave his bunk. Eventually, Mike arranged transport home via one of the passing steamers. Twenty years distant, I wonder if it wasn't in the dark, driving the *Patricia Lee* in green circles, that I found the space to finally recognize the taxa that might claim me the way brown king crab had claimed Mike. During those quiet nights, were the lists I made in my journal – the names of western trees and their identifying characters – my first recognition that, for me, going to sea was less destination than crucible?

AT SEA, THE PLATFORMS WE REST UPON – a boat, a kayak, even an island – are, by definition, spatially limited. Returning the next afternoon from our day-long paddle out past Gifford Peninsula, Maggie wants to land on both Josephine Island and the small islet beside it that she calls Josephine Island Junior. Earlier this morning, we'd found a stretch of shore cleared of rocks – likely a historic canoe skid maintained by the members of the Tla'amin First Nation – but there is no such landing on Josephine Island. Urchin-covered boulders crowd its shoreline. Finally, we decide that Marc will nose the big kayak in close enough for Maggie to jump out and run up onto the shore, and I'll photograph her landing. As I record the evidence, Maggie standing proudly first on Josephine Island, and then on Josephine Island Junior, I wonder how much of her first sea journey my daughter will remember.

Islands – bits of seabound land – have a way of altering those species or individual biologists who linger on them. In evolution, the isolation of islands combines with chance to ferment species change. In BC, island wolves are smaller and less aggressive than mainland packs. Rather than pulling down deer and fighting off grizzlies, these wolves prey more on clams and barnacles.[4] On other islands, raspberries lose their thorns, birds abandon flight and iguanas learn to swim.[5] The ocean may be the birthplace of large-scale differences in body plans, but islands are the cradle of species diversity.

For biologists, the patterns of islands punch far above their total land mass. Charles Darwin and Alfred Russel Wallace – the men who first articulated natural selection as a driver of species change – were both obsessed by what they saw on

Hare Point Tide: A Stratified Reflection

Desolation Sound

Hare Point

Strait of Georgia

In the first days of July, bright colours of flower, fruit and leaf embolden terrestrial life rooted in har...

In the long, slanting light of a summer evening, silver-sided fish jump, harbour seals bob, a kingfisher rattles and the tide floods to full. Walk the tideline—here, vertical granite knuckles free of water twice a day. This is the land where manzanita and arbutus tangle the edge of dry Douglas-fir forest, where wild onion, sedum and brodiaea bloom atop crackly pillows of lichen and moss.

harbour seal floats, at home in the liquid membrane that forms a diffuse boundary between air and land

and sea life shows itself at lowest tide, caught in shallow pools or wash...

Studio Illustration: Hare Point Tide: A Stratified Reflection

s the land of great exchange, where exfoliating
te cliffs soften into ledges for the sea's cucumbers,
, and urchins, for it's scrabbling crab, huddled
els and clinging Fucus. The great flux washes
d out, reworking geography, boating murrelets
ding rivers of saline through narrow inlets.
gold light races before the dark, retreating
highest contour, to the unshadowed
ction. Feel the pull.

islands. Although I dreamed of its volcanic flanks often, I never once stepped ashore Semisopochnoi Island, the visual centrepoint of the *Patricia Lee's* fishing grounds. As I watch Maggie climb back into the kayak from Josephine Island Junior, I realize it was not outcrops of sand or coral or bedrock but floating boats that served as islands in my life.

RETURNING HOME from a sea journey has its own difficulties. Just ask Odysseus. Even when I hesitate to begin, I most often resist the end of any voyage. On our final morning at Hare Point, it's clear Marc and Maggie share my ambivalence. We are slow eating breakfast and even slower breaking camp. Once the dry bags are stowed, Maggie asks if we can circumnavigate Josephine Island one last time before we head for home. Even though we know the outgoing tide grows stronger with each passing moment, Marc and I agree.

It's not just tide, but wind too, that drags at our boats when we come out from the lee of Josephine Island and turn southeast. At first, I'm more puzzled than alarmed. Southeasterlies are not the winds I worry about here. But by the time we reach the elongate Cochrane Islands – less than a third of our way back to our launch site – the wind has grown into a full-throated roar. Bright sun still shines, but I can't feel its heat. Wine-dark sea runs through wind-braided sprays of white. Each time my kayak bounces down, water splashes my face. Salt trickles into my mouth. Squalls race toward us, dragging walls of friction across the water's surface.

In front of me, Marc and Maggie paddle steadily, the distance between our boats stretching as the combined force of tide and wind saps my shoulders. I see Marc's face, shadowed gray under his ball cap, turning back to look for me.

It's too far, too windy, to shout. I know Marc's worried about me, but he can't stop until we find some protection. Only momentum keeps our narrow boats from turning broadside into the rising waves. And Maggie's in his kayak. This is not the worst weather we've paddled in, but it's certainly the worst we've paddled with our daughter.

Without pausing the movement of my arms, I lean forward to see the chart. It looks like the cluster of three islands off the end of Coode Peninsula might form a windbreak. It'll mean backtracking if we can't slip through the tidal gap between the last island and the peninsula, but I find myself willing Marc to make this gamble.

He does. When I turn into the narrow cleft, the water instantly calms. Marc's already onshore with the hatch of the double open. Maggie's turned in her seat, watching for me. As I slide in next to them, Marc hands me our last sausage. Without waiting for a knife, I gnaw off a hunk. Maggie's eyes widen. In my family, I'm the reluctant meat eater amid the carnivores. But I want calories. I think I'm going to need them.

And I do. Returning takes three times as long as our paddle out. The pull in my shoulders solidifies into spears of pain, even as Maggie and Marc pull further and further away from me. In the final kilometre, the water boils greasy and yellow-tinged above the sandspit guarding the bay we left three days ago. The moment I clear these breakers and turn downwind, the relief is exquisite. Not dissimilar, I think, coasting in on a roller, to what I felt when Maggie finally squeezed out of me.

GOING TO SEA is one of our oldest stories; no one returns unchanged. What did I learn from the days I spent living aboard floating platforms of wood and steel and fibreglass? Definitely a love of light, a hunger to know the living world, especially its plants, and, most importantly, the understanding that I thirsted for intimacy, not just with people but with place itself. Life at sea bred within me a desire to be shaped in the same way I understood Mike to be shaped by the underwater contours of his fishing grounds, the snow-covered slopes of Semisopochnoi Island, the filter of pale light through gray cloud.

But such shaping only occurs when we allow ourselves to become entangled with – dependent upon and protective of – the lives of those who live in place. For Mike, brown king crab was the frozen product that paid the mortgage on the *Patricia Lee* and the ancestors of the crab that his son might need to fish. No one monitored the catch – not even me whose job it was to ensure that only legal-sized males stayed on board – more carefully than Mike. Each time we ran a string of pots, I sampled a percentage of the catch; from his perch in the wheelhouse, Mike watched every single crab that went into the processing chute. Both the deckhands and I learned to expect the sharp barks over the PA.

"Throw that small one back. *Now.*"

For me, going to sea – on a fishing boat, in the 18-foot kayak I paddled the length of the Baja Peninsula, or on the natural history cruise ship I finished the decade on – meant I lived as a guest, not as an inhabitant. Going to sea paid my bills, but I would never belong there. If I wanted to know the full possibility of place, I had to learn to love *my* rocky and intermontane world; to not just know but to become dependent upon the names I first memorized in the calm dark of a wheel watch: *Populus balsamifera* ssp. *trichocarpa, Pseudotsuga menziesii, Pinus ponderosa, Tsuga heterophylla.*

Today, paddling in the final few feet, I remember what it was that finally stopped me from going to sea. It was a deliberate imagining, the most overt commitment I've ever made, first hoped for here in Desolation Sound more than a decade ago, when Marc and I came alone. The result of which now stands, only a head shorter than her dad, waiting impatiently for me to come ashore.

4. The Route Finding of Lines

KALEB, TALL, QUIET, with a quick smile and considerate air, walks beside me, uphill out of the canyon of Petersen Creek. On either side of us, Saskatoon blooms white above balsamroot's yellow. It's the first day of May, the first week of my month-long field botany course. The rest of the class follows behind us – their chatter of plant names and gossip a counterpoint to the soft chittering of the yellow-rumped warblers above. Here, at the end of our first field foray together as a class, I'm surprisingly content, maybe a little smug.

It's not that we've travelled with any great speed: when we started, more than four hours ago, I'd planned for us to descend the entire four-kilometre length of Petersen Creek Park before looping back to the university. Instead, we were barely a kilometre down the trail when it was time to turn around. No, it's not the distance nor the speed but the *mode* of travel that's pleased me. The students have been walking as botanists: eyes opened to the repeating patterns of leaf and petal shape, colour and number that are the language of any flora. All it took was a cluster of shooting stars – each flower with brilliant magenta-coloured petals reflexed away from black and yellow outstretched stamens – for the students to fall on their knees.

Literally. All 12 of them. On their knees, head down, bum up, field journal and pencil outstretched in front of them, the shape of each student echoing the cone-like geometry of the shooting star flowers. What would an unsuspecting hiker have thought rounding a corner to such a view – perhaps a new type of yoga, done al fresco, built not from the cobra or the downward-facing dog but the "shooting star" asana. Perhaps not quite what Pliny had in mind when he named this group of flowers *Dodecatheon*, after the 12 Olympian gods.

But, no, not exercise, not prayer, just the thrilling sight – at least to me – of botany students drawing in the field.

When we lose nature, we don't lose it all at once. Most often, plants go first. Walk into any university lecture hall in North America and most students can name more corporate logos than native plant species. In 2007, the *Oxford Junior Dictionary* deleted "almond," "blackberry," and "crocus," to make space for "analogue," "block graph," and "celebrity."[1] Most university students are largely oblivious to the plant world. The success of any plant course, and especially a taxonomy course like field botany, rests in the students losing their sense of the botanical world as an undifferentiated green blur. But once that green blur resolves into thousands of individual species, we need some way to make sense of this diversity. It can't be through rote memorization. BC is home to nearly 2,500 different flowering plants, and if this diversity registers nowhere near that of an equatorial landscape, it's still far too many for brute memorization.

Seeing, finding pattern: two different acts, both key to learning plants by name. Both helped, I've come to believe, through drawing. Convincing biology students that drawing belongs in science can be a hard sell, but artists have long understood the way drawing can transform *looking* into *seeing*. Yet any tool – pencil, hand lens – is only as good as our skill with it. In previous offerings of this course, some students have rightly grumbled that they lacked the skill to draw well. This year, rather than plunging full speed ahead into the characteristics that separate the flowers in the mint family from those in the figwort family, I made time for drawing lessons. It felt subversive, maybe sacrilegious, to stand in front of laboratory benches and a fume hood and lecture on foreshortening and vanishing points. But it seems to have worked. Or, at least, today the students didn't hesitate to reach for their pencils.

Nearing our vehicles, I ask Kaleb if I can see his journal page for our outing. It's good: a species list to the left, and an explosion of quick sketches and diagrams spilling across the middle, with natural history notes filling the right-hand margin. As I hand his field book back, Kaleb asks me when I started drawing.

It's a question I've been asked before, and the answer comes easily.

A sunny Saturday in early May. My 30th birthday. I'd just finished my master's degree and gotten my first job as a botanist working in Missoula, Montana, of all places. I'd arrived the day before and woken on my birthday, a little scared and

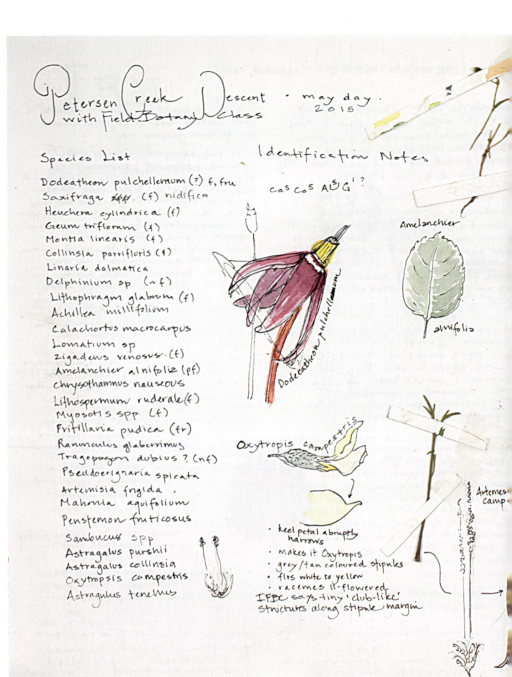

Petersen Creek Descent · may day. 2015
with Field Botany Class

Species List

Dodecatheon pulchellemum (?) f, fru
Saxifraga ~~xxx~~. (f) nidifica
Heuchera cylindrica (f)
Geum triflorum (f)
Montia linearis (f)
Collinsia parvifloris (f)
Linaria dalmatica
Delphinium sp (~f)
Lithophragm glabrum (f)
Achillea millifolium
Calachortus macrocarpus
Lomatium sp
Zigadeus venosus. (f)
Amelanchier alnifolia (pf)
Chrysothamnus nauseous
Lithospermum ruderale (f)
Myosotis spp. (f)
Fritillaria pudica (fr)
Ranunculus glaberrimus
Tragopogon dubius ? (nf)
Pseudoeignaria spicata
Artemisia frigida .
Mahonia aquifolium
Penstemon fruticosus
Sambucus spp
Astragalus purshii
Astragalus collinsia
Oxytropis campestris
Astragulus tenelius

Identification Notes

Ca5 Co5 AUG1 ?

Dodecatheon pulchellemum

Amelanchier

alnifolia

Oxytropis campestris

· keel petal abruptly
  narrows
· makes it Oxytropis
· grey/tan coloured stipules
· flrs white to yellow
· racemes ıl-flowered
IFBC says tiny 'club-like'
structures along stipule margin

Artemes
camp

essed plant for reference

MONTIA LINEARIS

PORTULACEAE!

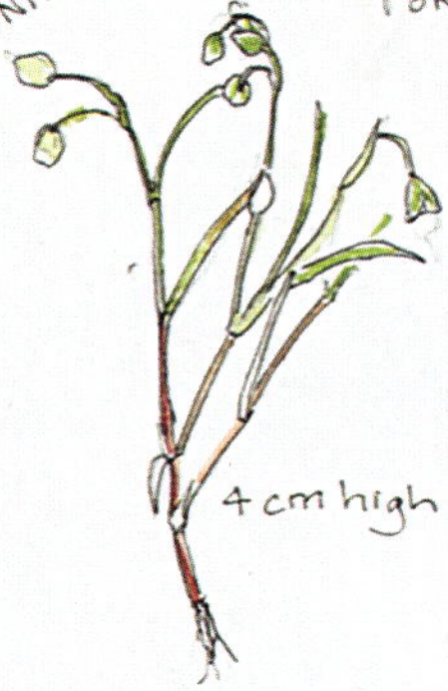

hrubby penstemon

4 cm high

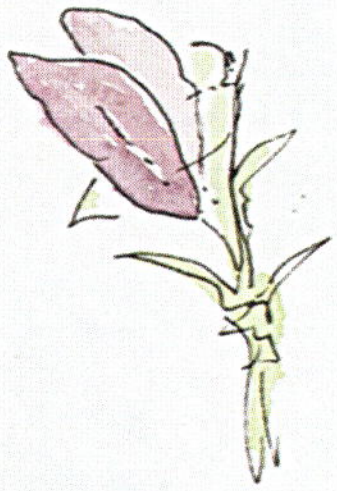

Other Notes

- north—ne facing side of creek gully
- mixed grassland, open doug-fir forest
- large amt of water dist. / erosion
- songbird activity
- active human use

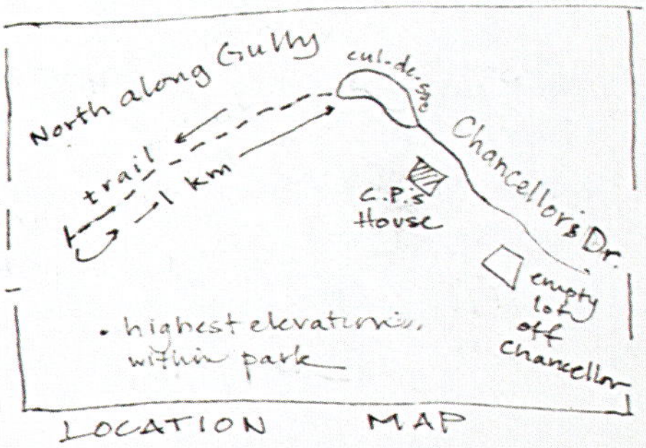

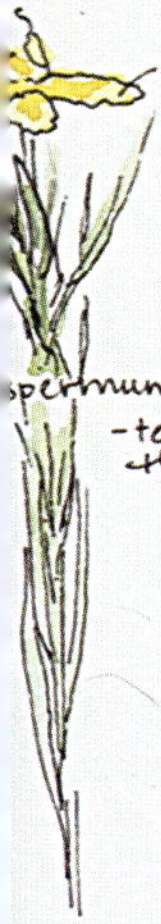

cadmium yellow! lobes ragged

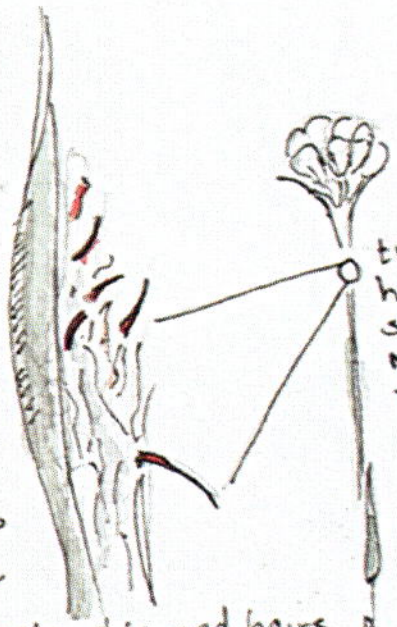

tiny red hairs on stem top; Not gland-tipped

incisum spermum ruderale
- tends to be smaller than L. ruderale

gland tipped hairs not found - just red hairs

woolly both top + bottom of leaf

North along Gully
cul-de-sac
trail ← 1 km →
C.P.'s House
Chancellors Dr.
empty lot off Chancellor

- highest elevations within park

LOCATION MAP

ANTENNARIA MICROPHYLLA

- bisexual flrs
- leaf shape
- presence of stolons
- mat-forming
- woolly

leaves brown at bottom

lonely in my new boss's guest room. I'd wanted to spend the day well and decided that if I was going to be paid as a botanist, I should be able to draw the plants I'd be naming. I'd gone to the University of Montana bookstore, bought supplies and spent the day drawing flowers on Mount Sentinel, an open hillside behind the university campus.

It's a good story. For those students resistant to the idea of drawing, it places sketching firmly on the side of skills that might help in getting a job. For those who haven't drawn since they got rid of their crayons, it's a reminder that drawing is a skill that can be learned.

Today, my story seems even more relevant. "In fact," I tell Kaleb, "the first flower I ever drew was the shooting star, *Dodecatheon pulchellum*, we saw earlier. I'll find the sketch and bring it in."

IT'S A GOOD STORY, BUT IT'S WRONG.

It's not that I didn't spend my birthday sketching on the flanks of Mount Sentinel – I did. I even find the sketch of *Dodecatheon pulchellum*. But I definitely misremembered its context. My drawings from that day are not the first in the sketchbook. There are others – a yucca from the Canyonlands of Utah, a heart-leaf arnica from Colorado's west slope, an *Iris* capsule from El Capitan meadow in Yosemite National Park. Some of them are not bad. None of them are the work of a novice.

If I'm wrong about *when* I started drawing, am I also wrong about *why* I draw? About *why* I ask my students to draw?

Maybe I'm just off by a year or two. Before long I'm surrounded by journals – one dating from my teens that I can barely stand to open, several from my university years, and a stack from my 20s. Most are pictureless: filled with text loopy or thin, written in blue, black, green, even red pen.

And then my eye catches sight of the yellow field books – four spiral Rite-in-the-Rain notebooks – that I kept as a master's student in the University of Vermont's Field Naturalist Program. I reach for them, the feel of their slick, waterproof paper carrying with it a raft of memories. Embedded within the Botany Department, the Field Naturalist Program was my first systematic exposure to a region's flora. Designed to give students the tools to read natural landscapes as if they were text, coursework in the program included a hands-on field practicum. My first year in the program, I spent each Friday visiting sites throughout northern Vermont.

Unlike reading books, reading landscapes used the whole body. We dug soil pits, separated rabbit from deer herbivory, cored trees, and made lists of flora and fauna. That's what I remember most: the ever-growing, omnipresent lists. Especially of plants.

Species after species. Name after name. Plant after plant. Late in the afternoon was always the hardest. Over the course of a long field day, the names would pile up, one upon the other, until their collective weight in my consciousness made it nearly impossible to hear one more name, one more set of identifying characters. Not plant blindness – plant fatigue.

Today, I take some comfort in the fact that this sort of mental fatigue is a well-known motivation behind any taxonomy. At their most basic, taxonomies group like with like, so that it becomes easier to focus on what is different among the individuals in a single group. Saying "a sunflower is a type of a flower, not a type of moss," lets us ignore the parts of the sunflower that separate it from all mosses (vascular tissue, seeds, flowers), and lets us focus on what distinguishes sunflowers from other flowers. The characters by which plants are grouped – size, use as food or medicine, evolutionary relationships – differ from one culture to the next, but all cultures classify. We have to. With nearly 400,000 plant species on Earth, the hierarchal order of a taxonomy serves as an intellectual crutch. It's the reason my master's program required a plant taxonomy course alongside its field practicum, the reason I have my field botany students memorizing the recognition characters of the 20 most species-rich flowering plant families in BC.

The first of my yellow Rite-in-the Rain field books is pictureless. But in the second volume – from a frigid winter ecology course in the heart of Maine – drawings appear. First, just rudimentary sketches of tree buds, but then, slowly, evidence of using drawing as a means of learning. By May, the drawing of sasparilla (*Aralia nudicaulis*) makes more sense than the words scrawled beside it. In September of my second year, rarely does a species get mentioned without an accompanying sketch. I don't remember being coached by my instructors. Maybe I'm misremembering again, but it seems as though when I was thrown into the details of a flora, drawing was a lifeline I reached for instinctively.

The last of these field books – the same book I found that contains the *Dodecatheon* sketch I remembered so well – is the most interesting. Not a Rite-in-the-Rain field book, but a spiral-bound artist's sketchbook. Used during my last field course in my master's program: a two-week botanical exploration of the Sonoran Desert. Sketches everywhere. Another flora being deciphered – this time with drawing an obvious handmaiden to learning. But then, at the very end, nearly 20 pages, each with a single illustration, each fundamentally different from those found earlier. The coral bean seedpod is an exercise in curves – the legume splitting and then arcing back together again. The wild cotton breaks through space via sharp angles and surprisingly straight lines. Not only do these sketches ignore standard diagnostic features, but, in noticing line and shadow, their contours invite me, years distant, to remember what in each plant first beckoned my attention. Some are left unnamed: drawing, not for learning; drawing, it seems, for drawing's sake.

What I think, sitting on the floor of my home office, journals stacked around me, is that if botany opened the door to drawing, then drawing, itself, opened me to a new way of experiencing the world. A way of experiencing that had less to do with the mental constructs of taxonomy than with hand using line and shadow to explore the lives of others. Unlike previous ones, these last drawings are dated. Twenty years ago, did I realize their importance, if only unconsciously? Twenty years: maybe my misremembering earlier today is part of the remaking of memory that occurs with time. Even so, that doesn't explain why for years I've identified my birthday, a year and a half later, as the date I started to draw. Why then?

I flip to the page immediately before the *Dodecatheon* sketch. A simple line drawing of *Iris* capsules in Yosemite National Park.

Yosemite. Home to ponderosa pine and pin oak, pine grosbeaks and an exfoliating pluton, named into fame: Sentinel, Half Dome, El Capitan. Never home to me, but during my 20s, an important destination. Without permanent address, I spent most of that decade – when not in graduate school or on boats – climbing high. Over time, I'd come to feel solidarity with the group of like-minded individuals who camped out for months at a time at climbing areas like Yosemite. A community that saw the world as a series of 150-foot pitches, and spent more time thinking about how to retrofit Toyota minivans into live-aboard campers than financing first homes.

Full membership in this community required leading my own climbs. Even as I'd been learning the names of plants, I'd been training my body to climb vertical rock faces. It hadn't been easy. I'd come to dread the self-doubt that could turn any vertical pitch into 150 feet of mental agony. The previous summer, climbing in Yosemite, I'd hiked up to the base of a technically easy, but extended, climb on an overhanging face called the Leaning Tower and walked away at the last moment, leaving the men I was climbing with to carry on without me.

On the same page as the iris sketch, I find an incomplete paragraph: "Above me on El Cap, two climbers and their yellow haul bag inch up the rock precipice. I am more grounded to..."

Grounded to what?

The answer to this unfinished sentence – one page before the sketch that I've long remembered as my *first* drawing – seems obvious. I'd gone climbing and ended up drawing plants.

In retrospect, drawing wasn't a bad substitute. Both artists and mountaineers are obsessed with lines. Climbers want good, clean lines in the rock that will let them summit safely; artists wish for bold lines that will let them draw true. Both practices require an embodied knowing, the muscle memory that can be acquired only through repetition. Both risk ruination – one undoubtedly a more mortal ruination than another, but each are only one misplaced line away from failure. When successful, both allow an intimate knowledge of a particular piece of the world – the crystalline touch of hard granite, the curve of a magenta petal – to reverberate through muscle and fibre, bone and ligament. Both practices alter their practitioners' views of the world: the view from high elevation can be as disconcerting as finding perspective in a drawing.

Artist and art critic John Berger writes, "Drawing is a form of probing. And the first generic impulse to draw derives from the human need to search, to plot points, to place things and to place oneself."[2] Running my finger overtop the sketch of the shooting star from Mount Sentinel drawn nearly 20 years ago, I trace the echo of that long-ago morning. I'd woken up resolved to spend the first day of a new decade well. Rather than climbing, I spent the morning drawing shooting stars.

My response to Kaleb's question, I realize, wasn't a misremembering. I merely answered a question different from the one he asked. My story explained not when I started drawing but when I first understood drawing as a form of route finding. That is, when I first began to understand that each time I bent to sketch a *Dodecatheon* flower or *Iris* capsule, I was gifting myself with the chance to imagine not just their lives but my own.

OVER THE NEXT MONTH, immersed in recognition characters and species identifications, I will be strangely reluctant to clarify the story I told Kaleb. As the students learn to separate the mint from the borage family, the aster from the carrot family, I will shy away from prompting them to think about what else, beyond identification, they might be encountering as they sketch sepal and petal, stamen and carpel. Drawing to learn is one thing; asking them to understand themselves and their place in the world through their pencils is another. But, as the students put their field taxonomy in increasing order, I can't help but wonder.

Later, near the end of our five-day camping trip throughout southwestern BC, our trailside chatter turns to the words that ornithologists use to refer to a collection of any one bird species: a murder of crows, a descent of woodpeckers, a party of jays.

"What," I wonder, "would you call a gaggle of botanists? A calyx, a corolla?"

"No," replies Kaleb. "Slow."

The laughter that erupts among these students is the laughter of the converted, the initiated.

Black lines on white paper. When we return home from our field trip, the students' illustrations are better than what I imagined possible. Much better. Cara's chocolate lily is built of bold lines, anthers peeking out from behind mottled petals; Dylan's *Viola glabella* is a detail-rich monster; Kayla's shooting star springs off the page; Matt's *Erigeron* sprawls; Isabel's yarrow gathers complexity; and Lucas's dimorphic *Antennaria* gleams in satisfied elegance.

The great wonder of this world. We can classify or climb it, but one way or another, in searching for it, we find ourselves. This year, the students have understood.

5. The Cost of Mobility

ANIMALS MOVE — it's our birthright, a gift from our ancestors in the form of duplicate genes that code for leg or wing or fin. But how willing or how far we go always bears the imprint of home. If where we dwell is patchy or ephemeral – say, a spring pond that dries each year – bodies sprout wings or retreat into dormant cysts that blow with the wind. If where we live is stable and continuous – a deep sea vent or an old-growth forest – limbs remain rudimentary and our wandering limited.[1] Most animals disperse only when young; a few continue to travel long distances throughout their life. Near or far, young or old – no mobility is without risk.

IT'S THE THIRD DAY OF 2005. In the murky light of a January morning, three of us – Marc, 2-year-old Maggie, and me – lean into the sharp curve of an entrance ramp, headed east out of Kamloops. From atop the Trans-Canada Highway, the view of this small city is not what you'd call pretty. To the north, a moving wall of westbound freight cars lumbers west, obscuring all but a narrow strip of snow-covered hills and low-hanging clouds. To the south, American fast-food outlets, low-slung motels, mini-storage buildings and car dealerships appear and then disappear, a sprawling geography of mobility.

Not for the first time, I think how happy I would be to leave this place – if I wasn't trying to call it home.

Part of it, I know, is just me. Mobility may be the evolutionary legacy of all animals, but, in the Anthropocene, no species travels more than humans. Each year, as our planet orbits the sun – a distance of nearly one billion kilometres – humans travel more than 20 times that distance across its surface.[2] Certainly, before Maggie was born, I'd put in my share of kilometres, following first school and then work, east and west, north and south, across the continent.

But part of it is Kamloops. All mobility has consequences. Migrating birds disperse plant seed beyond its native range; earthworms tunnel aeration into soil; salmon replenish forests with nitrogen carried from the sea. In the last 500 years, as human mobility has skyrocketed, so too has our impact. But constrained by geography or history, we've travelled some places more frequently than others. Kamloops, I think, is one of those places where our travelling footprint has fallen particularly hard. Geographically, this small city sits at a river confluence – its name originating from the Secwepemcstín word *Tk'emlúps*, meaning "meeting of the rivers." Today, two railways and one national and four provincial highways converge within the city, making it the transportation hub of southern interior BC. Mobility feels cemented into the very architecture of this place.

Two months of residence in Kamloops, and what do I know? This city sprawls outward – covering the same land area as BC's largest city, Vancouver, with less than one-sixth the population. Even our small neighbourhood, near the eastern edge of downtown, feels unsettled, divided by highway and railway, perforated by tunnels, paved with cement, dominated by a low-slung architecture that privileges quickness over quality. None of which makes me want to stay; all of it tempts a permanent getaway.

WHEN THE TATTERED EDGE of the city finally transitions into snow-covered fields sloping down to the cold gray of the South Thompson River, I feel my body relax. Outside, the air is leaden; snow, I fear, is imminent. But inside our van the three of us are warm and contained. This trip feels like the first good decision I've made in months.

When the phone rang a week ago, and the familiarity of a voice echoed through time and space, my response had been immediate.

"We're throwing Tracey a surprise birthday party. Will you come?"

Tracey Morland. My best friend during the last year of my Canadian childhood. The first friend I could bring into my mother's house without concern. It's her mom, Jackie, on the phone. And, of course, I said yes.

Beside me, at the wheel, Marc asks, "Does the South Thompson empty into the Fraser or the Columbia? I pull out our new map book. This question isn't unexpected. My husband charts any new landscape through the rise and fall of its rivers. What's new is his expectation I will have the answers.

Last November, our move from northwestern Montana to BC – Marc driving a 21-foot U-Haul truck, me following behind in our van with Maggie and our two dogs, cast me as the expert in more than just river drainages. BC is headed into a provincial election, and there is talk that a federal election might occur next year. News junkie that he is, Marc's been exploring the political terrain of his new home via CBC Radio. "What's a minority government?" he wants to know. "What's a vote of nonconfidence?" Given that I just spent the last 27 years living as an expatriate in the United States, it doesn't take much to deplete my understanding of Canadian politics.

This is what I *do* know: be careful about what you start. Last spring, the advertisement for what is now my job had seemed perfect. A position teaching university botany and ecology less than two hours from Irish Creek in southern BC. Not just a job, but a chance to return to the landscape I'd long held in memory as the heart of my home.

In my interview, when asked if I was prepared to teach the courses assigned to the position, I'd responded, "I don't think you ever *know* anything until you teach it."

They'd liked that. Nodded their heads. And their ready agreement let me avoid explaining that I'd never taken several of the courses I'd be expected to teach. That day, in the interview, I'd thought myself clever; in reality, I'd been at best, glib, and at worst, arrogant. Now, two months into the reality of my new home, I'd give nearly anything to skedaddle. Pack our bags, grab the dogs and the kid, and just leave.

RUNNING TOWARD; RUNNING AWAY: two different reasons to pick up and go. In my life, this latest move from Helena, Montana, to Kamloops – a distance of just over 1000 kilometres – has been unusual only in that its accompanying grief has been so hard to shed. Months ago, when Marc and I debated leaving Montana, I thought I knew how to calculate the risks and opportunities of a mobile life; when to stay and when to go. But for the first time I can remember, I have work that won't end, and I live in a house I have no need to sell. And, still, all I think about is leaving.

Obviously, I miscalculated. Badly. Looking back, it's easy to see that renovating our house and giving birth to Maggie in Helena wedded me to community in ways I'd not understood. My prenatal yoga class transformed into a circle of friends.

Neighbours monitored my pregnancy and ran unfinished errands when Maggie arrived earlier than expected. Here in Kamloops, I'm lonely and exhausted. I have lectures to organize, labs to write, grants to propose. I live with a constant, if low-grade, case of diarrhea. Failure, I fear, is imminent. Last week, I started imagining scenarios that could let me leave without insulting those who hired me. An allergic reaction to the pulp mill in town? A death in the family?

I can't tell if the true kernel of my despair is that I feel incapable of resisting mobility's pull, or that we've nowhere to go. We sold the Craftsman bungalow we'd just finished restoring in Helena. Marc quit his job as a wetlands ecologist – work he loved – so I could take this one. Last night, driving home, the lack of an obvious escape reduced me to tears. Seven minutes of solitary grief in the winter dark, before I pulled up in front of our new blue bungalow and rubbed my eyes dry.

Today's mobility is, at best, temporary.

OUT THE VAN WINDOW, crooked cottonwood trees back up against the South Thompson River. Each one of these trees – organisms without muscle or bone or neurons – once depended upon the mobility of seeds. Released just as flood waters are beginning to subside, cottonwood seeds are tiny – only 2–3 millimetres long – and held aloft by a cotton of silky hairs. Snagged by shrubs or branches, many will never touch the ground; others pile in miniature drifts along road or field edge. Sculpted by evolution, the mobility of these trees is a one-time event, an all-or-nothing gamble on the whim of wind and flooding river waters.

Just before our route turns overland, I wonder when the seeds of these trees will next disperse. Sometime in May or June, I assume, but I'm not sure how the change in latitude and longitude (north four degrees, west eight degrees) and our loss of elevation (nearly 1000 metres) will influence the phenology of these cottonwoods.

When we turn south, driving up and out of the South Thompson River valley, I need to use our map book to give Marc directions. But the longer we travel, the more I recognize. In 1971, my hippie family wasn't the first of our kind to arrive in Armstrong – a small farming community in the North Okanagan Valley – but we were close, arriving with what could be packed into a 1954 Studebaker. My family may have gone back-to-the-land, but we lived like our annual crops: dependent upon the soil but never fully settling in place. Instead, during the six years we lived in the North Okanagan, we regularly moved from one rented house, one part of the valley, to another. Our last move, after my mother and second husband, Simon, separated, brought us into town. That rented house was a converted garage, but unlike most of our previous rentals, it had electricity, running water, a phone.

Of all our new amenities, the phone – matte black with a rotary dial, hanging on the kitchen wall – was the best. A form of bodiless mobility, voices weaving me into a larger after-school conversation. Little did I know that, before the school year was finished, this same phone would announce my family's next move – this time across an international border.

That had been in January too. Right after the only Christmas holiday my siblings and I ever spent with my biological father. On Boxing Day, Mom had driven us to the closest airport. When my siblings and I boarded a plane – our first ever – to Vancouver, Mom got on a plane to visit her new boyfriend, Glenn, in Montana.

On New Year's Day, she'd called us. She had a surprise, she said, that she'd tell us when we got home. Maybe, I thought, Glenn, who was quiet and read books, was coming to stay. When we were back home, and the phone rang, I raced to get it, Mom's voice scratchy and distant.

The surprise? She'd gotten married. No, not to Glenn, but to a man named Charlie. The last thing I heard before I handed the phone to my sister was, "We're moving to Montana."

The winter night my mother called to tell us she'd married a stranger, I didn't stay long enough to hear my sister's reaction. The moment Laurie took the phone, I grabbed my coat and ran toward the Morlands' house.

Theirs was another hippie house. When Tracey and I met in Grade 5, I'd gone through the hippie-kid ritual of parsing any paraphernalia I found in her house. Bead curtains hanging in doorways; macramé plant hangers crowding windows. Finally, I'd copied the lyrics to Dr. Hook's "Freakin' at the Freaker's Ball" and handed the words to Tracey as we walked home from school. It was enough. On Christmas Day, Tracey had brought her family over to my house, and the adults got stoned before we all went sledding.

That night in January, Donnie, Tracey's dad, came upstairs when he heard my entrance, and the three Morlands sat with me at their square kitchen table as I sobbed out my story. After I'd calmed down, Donnie, in good hippie fashion, wondered if I shouldn't consider this new move not a tragedy but an adventure.

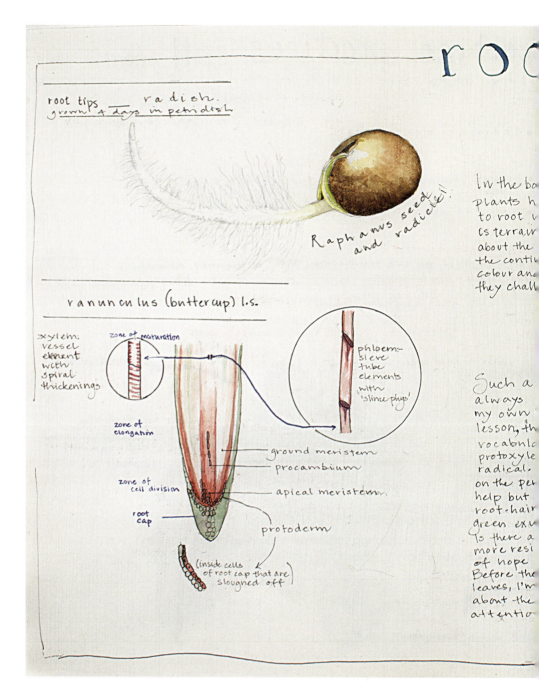

In the bo...
plants h...
to root...
is terrain...
about the...
the contin...
colour and...
they chall...

Such a...
always...
my own...
lesson, th...
vocabul...
protoxyle...
radical...
on the per...
help but...
root-hair...
green exu...
Is there a...
more resi...
of hope...
Before the...
leaves, I'm...
about the...
attentio...

Vol. 10: Roots

in the Botany Lab S364 for BIOL 221
TRU Campus

_____

ture roots, prepared slide

nunculus x-s of root

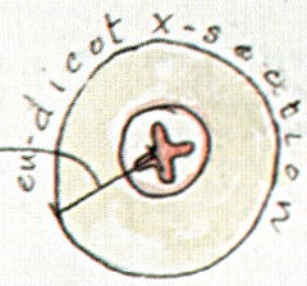

eu-dicot x-section

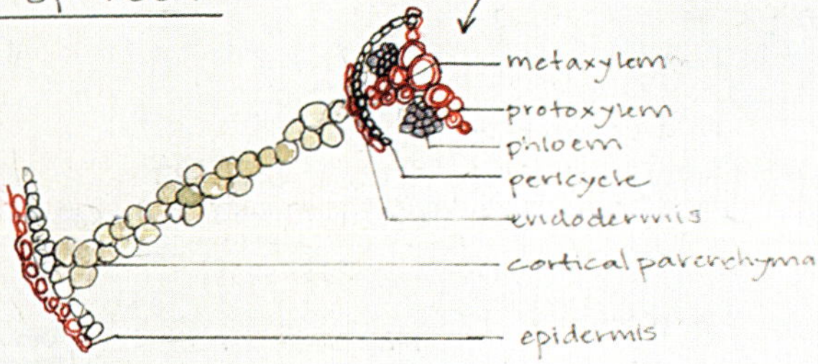

— metaxylem
— protoxylem
— phloem
— pericycle
— endodermis
— cortical parenchyma

— epidermis

s clear that
naw one- way
of this anatomy
, but I think
ilar labs across
ling the shape and
plant rooting. Are
, to think beyond animal mobility?

_____

ocot x- section of root
ea mays, prepared slide

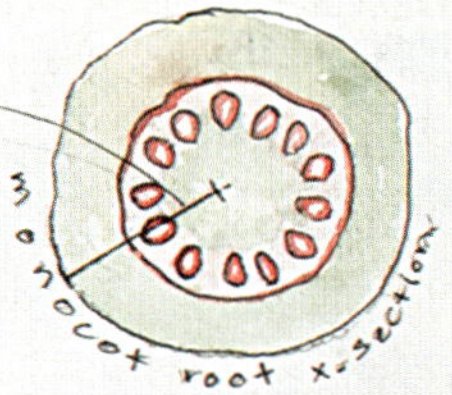

_____

o find buttercups (Ranunculus) —
pendable point of yellow in
in integral part this rooting
ecimen to specimen. The
ctinostele and atactostele;
xylem; root hair and
that relies so heavily
fixed slides, I can't
npossibility of the
dicle, the bap
two cotyledons.
,insistent,
e container
ds?
,t
dering
,lp between
thy; empathy and change.

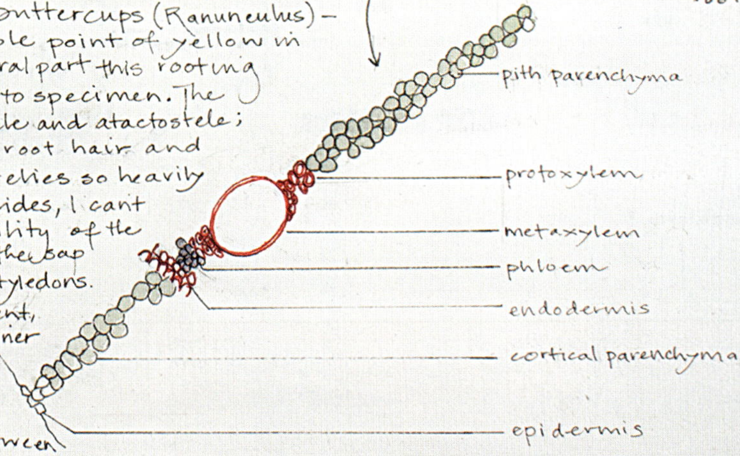

onocot root x-section

— pith parenchyma

— protoxylem

— metaxylem

— phloem

— endodermis

— cortical parenchyma

— epidermis

Mobility, dispersal; opportunity, risk. Good or bad, you trade familiar for strange, known for unknown. When my family left BC in 1978, it was easy to list my losses. The only home I could remember. The kids I'd gone to school with since Grade 1. The comfort of a best friend with hippie parents. But it took me years to understand how one move can lead to others.

As a plant biologist, I chose to study a group of organisms who root in place. But I also chose a profession – university teaching – that assumes, even rewards, long-distance mobility. In his essay "The Rootless Professors," Eric Zencey argues that a transient class of intellectuals, who owe no allegiance to a geographical territory, largely provides post-secondary education.[3] Against all odds, joining the professoriate allowed me to return to the landscape from which I was exiled 27 years ago. But, in our van, driving to the Morlands, I wonder about my allegiances. Do I have any real loyalty to the plants of this *place*: this valley, this watershed, this country? Or is my loyalty more to the *ideas*, the *practice*, of my discipline?

This is what I want to know. Do those cottonwood seeds, who fail in dispersal – piled in drifts, suspended in the air – mourn the roots they will never know? Do the proteins in seed coats know their thirst as they reach for water – the first step in germination? Is the risk of mobility easier if you never have a choice, if you go by chance, or by the desire of another? Or is the true cost only known when it's you who makes the decision when to go, when to stay?

IN THE VAN, Marc wants to know if we should turn left onto Salmon River Road. "Yes," I say, already opening the map book, expecting his next question.

"So does the Salmon flow north into the Thompson, or south out the Okanagan?"

Before I can decipher the blue lines on the map, there are more directions to give. I guide him onto Deep Creek Road, and then another left and a right, and then there's the Morlands' house. Emerging from the side of a hill, the low-slung wooden house – built after my family had already left – is not one I know. The Morlands may not have been the first hippies to arrive in this valley, but unlike us, they *stayed*.

In the life of cottonwoods, dispersal is easy; establishment is more difficult. Within 24 hours, seeds that land on sunny, moist riverbanks will split open to produce a sticky-haired foot. Once the foot stabilizes the seed, the embryonic root, the radicle, expands, pushing down into the soil. And then, over the next few weeks, it's a race. To maintain contact with receding flood waters, root tips must expand as much as a centimetre each day. Bank position matters. Too high on a stream bank, and juvenile roots will wither from thirst. Too low, and stems will be scoured by next winter's ice, or drown in spring's flood. Of the seeds that disperse from a tree, only one in a million successfully roots.[4]

IN THE ENTRYWAY OF THE MORLANDS', it's a bustle of hugs and hellos. Donnie and Jackie are older versions of the gracious hosts I remember. As we settle into the living room – woodstove radiating heat, windows overlooking snow-covered lawn – I sense Donnie trying to read Marc's disposition. Refreshments differ in kind and impact; it's important to get it right. Marc's short, nearly crewcut hair, full beard, worn jeans and faded work shirt is the attire of a field ecologist, but collectively the different pieces have left Donnie guessing. Finally, Donnie errs on the side of ambiguity.

"Marc," he asks, "how can I alter your consciousness?"

Marc looks stunned, and I grin. My American husband has nothing against a finely malted scotch or a good porter, but he's generally comfortable with his consciousness just as it is. It's also mid-morning, and we have to drive back through the now-falling snow to Kamloops. I can see Marc wondering where the boundaries of this particular culture-shed are.

"Coffee will be plenty," is his considered reply.

It's a good party. There are a few faces I recognize and more that I'm told I once knew. When it's nearly over, I hear Donnie's voice boom out from the kitchen, "Well, maybe I threw away my vote when I supported the Marijuana Party in the last election. Don't know that I'll do that again."

Marc, standing nearby, asks, "A Marijuana Party? Did it run candidates province-wide, or just here in the Okanagan?"

I'm disappointed not to hear more when Maggie calls me down to her level.

But on the way home, Marc only has to say, "The Marijuana Party? Really?" before we erupt into giggles.

Welcome to my history, babe.

IN THE VAN, we are back alongside the Salmon River. I reach for the map book. From here, the Salmon River flows north into Shuswap Lake, which in turn is drained by the South Thompson River. Tracing the blue lines on the map, I realize there's a divide – a barely distinguishable uptick in elevation – between here and my childhood home. I thought moving to Kamloops would be a homecoming. But I was wrong – both geographically and culturally. The home I was exiled from as a child drained south down the Okanagan Valley into the Columbia River. The terrain of our new house, my new job, drains west into the Fraser River, before spilling into the Pacific. And, culturally? I might carry a Canadian passport and two graduate degrees in botany, but I am like a cottonwood seed caught in a shrub – isolated by my time away, by my ignorance of Canada's most basic civil traditions, by my lack of intimacy with this *place*.

BACK ALONG THE SOUTH THOMPSON, cottonwood trees braid twig into branch, branch into main stem. I can't see it, but below ground, cottonwood roots reverse the process, branching from main stem into increasingly smaller tributaries. Trees are mirrored rivers, their form collecting from both sky and earth.

But only if they remain in place. Once rooting occurs, further mobility risks everything. Large plants like trees get transplanted only with extreme intervention. To move a tree, a gardener must trench through the outer periphery of the tree's roots months before the move. The trench, in isolating the tree's roots from its extended community, forces the tree to grow new feeder roots in a much smaller root ball – a truncated form of rooting that can be moved, albeit with front-end loaders and cranes.

Even then, there will be scars.

In our animal mobility, we humans assume we can bounce around, forsaking community; calling one place home, the other not; going back-to-the-land without learning the plants native to its soil; relying on generalizations to make sense of the world's vastness. But mobile doesn't have to mean *rootless*. As we drive into the eastern edge of Kamloops's sprawl, it occurs to me that my grief may reflect not failure but success. I may think of myself as a transient professional, a rolling seed, but isn't my persistent sorrow a sign I'd finally begun to root in Montana? As we pull up in front of the bright blue bungalow on Pine Street that we paid too much for, and that I have yet to love, I can feel the first trickle of grief slipping into question. What, I wonder, would it take to be as rooted as a tree?

Rooting, I know, rarely occurs alone. Below ground, embryonic radicles survive best in relationship. Carbon-rich molecules drip, squeeze and pass outside root cell walls, and in doing so, they feed microbial multitudes. In return, root growth is modified, regulated, defended, orchestrated. In their rootedness, few trees can abandon community and survive. As I open the back door of our van, unbuckle Maggie's car seat and feel her sleepy body settle into my arms, this is what I think – beginnings matter. Even if they're hard to articulate, even if they consist of more question than answer.

What would it take to be as rooted as a tree? Years later, driving up to this same blue bungalow and worrying about how close we came to leaving, I will remember this question as the first sticky foot of rooting. In rivers and trees, it is the passage of water, slipping from one pore to another, coalescing into torrents, that links earth with sky, mountain with ocean. In the reciprocal relationship between people and place, stories do the same.

This was the first.

6. Community Matters

IT'S EARLY MORNING, mid-June, in the Lac du Bois grasslands just north of Kamloops, not far from the small waterbody that is unnamed on the map, but that I will come to know by the discipline that first brought me and many students here – Botany Pond.

On hands and knees, Laura, a new research student, and I bend over a quadrat, isolating a grassland plant community into samples. Many plants are known; some are not. Eventually, we'll need to find the Latin names for all community members, but for now we make up names for the ones we don't know. *Little red stem. Whorled leaf. Big white.* Laura wears our growing list of unknown plants, collected into individual Ziploc baggies, from a D-ring clipped onto her belt. Each unknown plant, roots and all, has been carefully extricated from the soil, cleaned of debris, then bagged and labelled with its temporary name. At the same time, just in case we lose the specimen, we list the plant's temporary name, its physical description and original location into our Rite-in-the-Rain field book. Check, cross-check.

In the field, Laura's collection serves as a walking library, easily flipped through for reference as we move from one quadrat to another. In the lab, preserved in a plant press, Laura's collection will serve as a botanical Rosetta Stone, linking one day's collecting with the next. I don't remember the day I learned Laura's name. I do remember how, last fall in my second-year botany lecture, I first recognized and then grew to depend upon her correct, even insightful, answers to my questions. In encouraging Laura to apply for the research scholarship that is paying her summer salary, I had pegged this quiet, long-boned, dark-haired woman as a natural candidate for fieldwork, yet the more time we spend together in the field, the more she has resisted easy classification.

Unlike me, Laura has stayed in place for university. From her childhood home, less than ten kilometres as the crow flies from where we now sit, Laura and her dad have explored, on mountain bike or foot, many of the trails that wind through the Lac du Bois Grasslands Protected Area. For her, this grassland is well-travelled terrain. Not for me. After a rocky first year teaching botany to mostly less-than-enthusiastic university students, today's field sampling is work whose contours I'm grateful to recognize.

Implicit in our sampling, as is true with most ecology, is the assumption that the right way to know community is to break it down into smaller, more manageable chunks. Earlier this morning, Laura and I used compass and field tape to outline a rectangular plot – ten metres wide, 25 metres long – around one randomly located point in the grasslands. The borders of our plot are not perfectly straight – robust fescue, lichen-covered boulders introduce slight hesitations, minor detours in the white tape – but by pinning the four corners down with tent pegs, we've got it pretty close.

Against the grassland's green exuberance, our plot is a transient hieroglyphic of estimation, festooned at irregular intervals with bright orange survey flags. Even in this relatively small space – not much larger than the yard that surrounds my family's new bright blue bungalow – it's difficult to accurately estimate the abundance of a single species. Instead, Laura and I place a small quadrat built of PVC pipe on the ground next to each orange flag. Peering down from above, one of us calls out how much of the quadrat is physically occupied by each plant species, while the other records the information on our data forms. We don't dither over the exact percentage – estimating abundance on a scale of 1–6 gives us enough accuracy.

Laura and I have been climbing a steep learning curve, deciphering community, one species, one quadrat, at a time. We've had to re-jig our data form to make room for the 70–80 species that inhabit any one plot. At the beginning of the week, we were clumsy, having to translate our sampling protocol into step-by-step directions for our bodies. But, this morning, we've not only remembered all the field equipment we need – 50-metre tapes, tent pegs, GPS unit, clinometers, field guides, compass, data form, clipboards, Sharpies, pencils, quadrats – but through the sheer weight of practice, we've sorted out who carries what and which vest pocket works best for what gear. Even our conversation has diminished; both

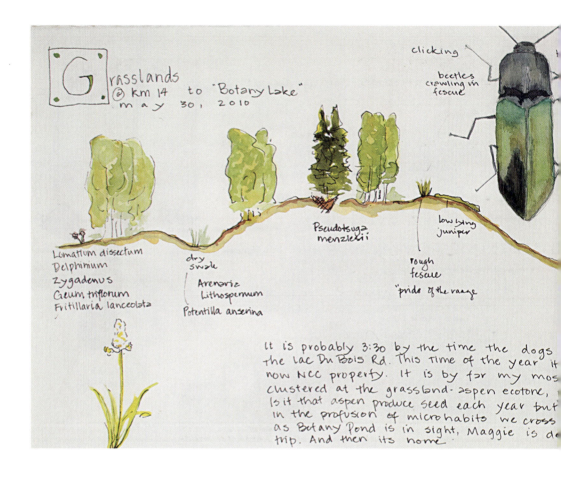

The handwritten journal illustration reads:

Grasslands
@ km 14 to "Botany Lake"
may 30, 2010

clicking

beetles
crawling in
fescue

Pseudotsuga
menziesii

low lying
juniper

Lomatium dissectum
Delphinum
Zygadenus
Geum triflorum
Fritillaria lanceolata

dry
swale

Arenaria
Lithospermum
Potentilla anserina

rough
fescue

"pride of the range"

It is probably 3:30 by the time the dogs
the Lac Du Bois Rd. This time of the year it
now NCC property. It is by far my mos
clustered at the grassland-aspen ecotone,
Is it that aspen produce seed each year but
in the profusion of microhabits we cross
as Botany Pond is in sight, Maggie is de
trip. And then its home.

Vol. 16: Grasslands @ km 14 to "Botany Lake"

Laura and I slipping into the absentminded call-and-response of plant sampling that leaves ample room for mental wool gathering. "Little red stem, cover class 5. *Festuca campestris*, 3."

Community matters. It could be a mantra for my career as a field botanist. But "community" is also one of those slippery words ecologists use without necessarily agreeing on what they're talking about. Two years ago, in preparation for my PhD defence, I memorized one textbook definition of community as "the species that occur together in space and time."[1] Pretty terse – maybe because that's the limit of agreement between ecologists.

Anemone multifida
pink phase!

ne multifida!
both white + pink
orphs!)

and Maggie and I clamber out of the van at Kilometer 14 on
flagrant negligence if I dont squeeze in one trip a week to the old Froleck-
spot in the park! Chocolate lilies are in explosion. Masses of them
the road. There is fluff in the air – "aspen fluff". I am confused.
don't germinate. all the old favorite flowers are in bloom and I delight
aphy contributing to the diversity! Once "Botany Lake" – probably better
look for aquatic creatures – leftover enthusiasm from the McQueen Lake

Part of the problem is that how we understand community depends, in part, on how we break it down. Long, skinny plots find more rare species than short, fat plots. Plots oriented parallel to contour lines count fewer species than those oriented perpendicular. Today, we know enough to understand how plot design – really, a series of rules we use to define the territory of our attention – can influence what we describe as community. What is still debated, however, is what it means to live in community.

Grow any two individuals next to one another, and, by definition, they will interact – even if it's only through sharing the same space. We know that, given enough time, species can adapt to one another, resisting or accommodating one another's existence. What is less clear is the influence these interactions have on which species "occur together in time and space." Are the relationships between neighbours strong enough that when you find one neighbour, you always find the rest of the community? Or is the presence of individual species determined less by their relationship with neighbours and more by their individual tolerances of the light, water or nutrients found in a particular neighbourhood?

In ecology, this has not been a trivial debate: being on the wrong side of prevailing opinion banished at least one man from the discipline. In the first decades of the 20th century, the majority of North American ecologists, led by Frederic Clements, a strong-willed, some might even say dominating, ecologist from Nebraska, believed that communities could be best described as something akin to a superorganism – highly organized entities made up of mutually dependent species. In 1926, when Henry Gleason, a young, upstart ecologist from Illinois, published the first in a series of papers arguing that communities were best thought of as a "coincidental assemblage of species," he was so shunned by other ecologists that he became, in his own words, an "ecological outlaw, sometimes referred to as 'a good man gone wrong.'"[2]

Like most North American ecologists working today, I've been raised on a more Gleasonian notion of community. In fact, the plot Laura and I are using today, a *modified-Whittaker plot*, is a variant of the design Robert Whittaker used in the 1950s to collect the data that ultimately shifted ecology toward Gleason's ideas – long after Gleason had abandoned ecology. This winter, when Laura and I were finalizing our field methods, our choice to use the modified-Whittaker plot design felt weighted on the right side of community.

But today it's clear that the data Laura and I will collect in any one plot is not, by itself, community. At best, it's a translation of community from the green tangle beneath our feet into the dispassionate numbers and graphs of science. It's a translation that, by design, says little about the immeasurable syllables of experience – the sight of white clouds scudding through blue sky, the itchy give-and-take of a mosquito finding the exposed skin on my hands, the feel of a hard-edged boulder imprinted on my backside – that collectively write another translation of community. Right now, here in the grassland, as birds flit overhead, butterflies land on nearby flowers and Laura calls out plant names, I feel the bubbles of something – contentment, joy, maybe even the first stirrings of love –

burble up from deep inside me. How we *feel* about community, I realize, is an important complication in how we *learn* community.

All week, I've been happy to decipher the intimate details of one "coincidental assemblage of plants," even while resisting any knowledge of another. My family's new house lies near the valley bottom, more than 500 metres lower in elevation than this grassland. Its previous owner seems to have been an enthusiastic, if somewhat chaotic, gardener. This spring, a diverse array of species planted with intent – irises and daylilies and an unusual ginkgo tree – have erupted into leaf only to struggle against an equally diverse assemblage of weeds. Weeds that need pulling; weeds I've been surprisingly content to ignore.

The species in my garden "occur together in time and space," but I haven't made the time to learn their version of community. Maybe it's a form of vegetative elitism that favours first comers over latecomers. Ecologists arbitrarily distinguish between natives – those species present before European settlement – and exotics – those species that came with or after the first European settlers. Most of the species in my garden are recent transplants, their continuing presence largely dependent upon the good will of gardeners. But I don't think it's just snobbery. Technically, all the plants in this valley came from somewhere else. Seventeen thousand years ago, when glacial ice layered more than a kilometre above our heads, this land was bereft of plants. Post-glacially, we are all migrants; some merely came later than others.

It's time to move again. Laura scrambles up to her feet and leads the way to the next quadrat. I follow, thinking about how the unknown becomes known. The collection bouncing at Laura's waist is one way. Back in the lab, we'll use technical keys and dissecting scopes to label each specimen with its Latin name. Later, Laura's thesis will organize our translation of community into a summary that can be shared with other ecologists. I can list the communities I've deciphered this way: the deciduous forests of Vermont, the high-elevation lodgepole pine forests of Colorado and Idaho, the temperate rainforest of Vancouver Island. What's not on this list, I realize, is the garden I left behind in Montana. I never once drew a graph for this garden, never once counted its plants, but there's no denying it was a coincidental assemblage of plants that I knew intimately, that I loved.

There it is – the real reason my new garden is languishing. I don't have the heart for the intimacy of weeding and planting. Not yet. In whatever form they take, the communities we love evoke our loyalty, even when we desert them. As

Laura and I settle down in front of the next quadrat, I'm grateful it's my turn to call out names and abundances; hers to record. The possibility of finding new species will distract me from those I've abandoned. But, ugh. The first species I see in the quadrat is neither novel nor beloved. I've pulled this species up along roadsides, dug its long taproot out of my Montana garden. I knew it was here, but this is the first time we've found it in one of our plots.

It's not that it's ugly. My mother-in-law still hasn't escaped the ecological faux pas she made the first time she saw it growing by the roadside in western Montana and exclaimed, "Oh, what's that beautiful wildflower?" From a distance, with its bright purple flowers spilling over the top of spreading branches, spotted knapweed could be called pretty. Yet, for some ecologists, admiring its beauty is like exclaiming over the intricate armour worn by Mongol hordes galloping toward your small village.[3] In Montana, I watched this species crowd out its newfound native neighbours across an entire grassland slope, transforming a rainbow of floral pigment into a sea of purple.

Ecologists may not agree on what it means to live in community, but we're damn sure united when we think a species is living *outside* of community. It's one of the places in ecology you can hear emotion slipping through the objective constraints of science. Ecological descriptions of knapweed's behaviour in North America read like a military history. Knapweeds are described as "aggressive weeds [that] displace native species, change plant community structure, degrade or eliminate wildlife habitat, and reduce forage for livestock."[4] No official territory wants to be known as the beachhead that allowed its initial invasion. Technical reports written in Montana emphasize that knapweed was first collected in North America outside Victoria, BC, in the late 1800s – germinating unbidden from discarded soil used as ballast by ships crossing between Europe and North America. In BC, reports are quick to argue that the real knapweed invasion spread west from contaminated alfalfa seed introduced into Montana.[5]

No one wants it; everyone knows it's scarily robust. It's good at tolerating drought and grows well, even thrives, in disturbed soil. Worst of all, it reproduces faster than rabbits. In some habitats, spotted knapweed produces more than 10,000 seeds – the vast majority of which will germinate – in each square metre.[6] Hordes, indeed.

It's not that it can't live in community. In its native range on the grassland steppes of Eurasia, spotted knapweed is just another species among many – there you don't find it crowding out neighbours like it does in the soils of this continent. Since knapweed's introduction to North America, there has been an army of ecologists hell-bent on understanding its success. Slowly, but surely, the evidence is accumulating. *Community matters, but so too does history.* Place – and the interactions that occur between individuals within any given place – shapes who we are as individuals and species. And there are real consequences when individuals are dislocated from one community, one region, one continent, to another. Arriving in North America, spotted knapweed escaped from known predators and came bearing novel root chemicals unknown to native North American grasses. The impact of these novel chemicals has been compared to the re-curved bow that contributed to the Mongols' consistent victories over larger armies in the 13th century. Knapweed even appears to be able to manipulate relationships with North American soil organisms – fungi and bacteria – to favour its success over North American grasses and wildflowers.[7]

The one aspect of knapweed's ecology with which everyone agrees is that it was *accidentally introduced* to North America. Maybe it's because I still haven't recovered from our move from Montana – but for the first time, I perceive the particular slant contained within this two-word phrase. It's a conveniently anonymous and passive description that focuses more on the consequences of arrival than on the causes of departure. No knapweed plant ever set off to invade North America. The knapweed plant in my quadrat is a displaced species, a descendant of seeds abducted long ago from community. Introduced – by whom? The phrase ignores humans' role in knapweed's history, but I can't. Not anymore. Not with my history of unquestioned mobility. Not when I think of all the collections I've made across this continent.

It is, after all, only a matter of degree that separates knapweed from the plants bouncing at Laura's hip. I'm not so worried about the impact of collecting on individual species as I am about collecting's implicit assumption. Laura and I follow a set of elaborate protocols, used by botanical collectors worldwide, to mitigate our impact on plant populations: we make sure we can recognize rare or endangered species that grow in the grasslands; we never collect any specimen from a population with less than 20 individuals; and we work hard to minimize any disturbance to surrounding plants. We won't even take this collection far from native ground – only 20 kilometres down the hill to the university. But there's no denying that the individual plants we've collected have been removed from community. In collecting, I've always assumed the collections I've made have been a fair representation of community, never stopping to wonder what has been lost the moment I pull a plant from the ground.

Years ago, learning to make botanical collections gave me entrance into a scientific community rich with tradition and intent. There is no doubt in my mind that Laura has all the makings of an exceptional field botanist. Right now, sharing the lessons I learned from my botany mentors with Laura has been the only aspect of my new job that has felt right, but I know that this tradition – scientific collecting – favours my tendency to learn community by breaking it down into transportable bits more than it cultivates the patience necessary to experience it whole.

This is what I think, as Laura and I move to the last quadrat of the plot. Every collection is made in the service of something: scientific knowledge, desire, nation building, the sublime. What would a collection in service of *community* look like? Kneeling down on green grass for the last time today, I sense the tangles of experience that might come if my collecting worries less about ideas and more about *this* place: the red-naped sapsucker that will pop up, jack-in-the-box-like, from its cavity in an aspen tree; the cryptic green frog orchid I will find after years of missing it; the two sandhill cranes that will, on another June day, stalk me through tall grass; or the unknown predator whose deep tracks I will find in the snow on a solitary ski.

Nearly a century ago, the debate between Henry Gleason and Frederic Clements taught us there's more than one way to imagine community. Collecting plants in standardized plot designs has been, and will likely remain, an important step in my learning, but I can no longer ignore the hard-to-measure stories that integrate the entire community, exotic and native, human and more-than-human, into a whole. Knapweed teaches us there are consequences when the interactions that underwrite community are ignored. Down the hill, the plants in my garden remind me I can't discount the time required for these interactions to develop. If *community matters*, then I need to find new ways of collecting – ways that both teach me *about* and embed me *within* community, ways that acknowledge both the gift and the responsibility of membership. That, I think, would be a collection worth preserving.

7. Mapping Moss

I'M ON MY HANDS AND KNEES when I hear voices. From my perch atop a silty cliff rising from the east side of Okanagan Lake in southern BC, cottonwood leaves shimmer between me and the blue water below. Through gaps in this green curtain, I glimpse bare legs, feet in flip-flops. Two people, maybe more, headed for the beach at Three Mile Point. Nobody looks up, and I'm relieved.

Sometimes, it's just easier *not* to explain. Especially when I'm burdened with a miscellany of field gear: backpack, binoculars, hand lens, GPS, spray bottle, collecting pouch. Especially when I'm crawling along, nose pressed nearly to the ground, butt up in the air. Especially when I'm looking for a plant that can't be seen. Or at least not seen with the naked eye.

Finding what I'm looking for requires a willingness to plop down on hands and knees, squirt water onto silt, bring hand lens to eye and fall into the Lilliputian world of a biotic soil crust. Watered under a hand lens, the contorted contours of a dry soil crust will – before your very eyes – uncurl into a sparkling tapestry of green and gold. In BC's coastal forests, bryophytes – those small mosses and liverworts of the plant kingdom – hang from tree branches, clamber over downed logs and sprawl across the forest floor. But, in a biotic soil crust, mosses, along with lichens, *are* the trees, towering above a complex web of microscopic cyanobacteria, fungi and green algae. Within this community, the rare moss I'm looking for, *Crossidium seriatum*, forms a part of the sub-canopy, growing no taller than a millimetre in height.

If this moss is hard to see, it's nearly as difficult to identify. Distinguishing this species requires seeing the shape of microscopic bumps found on leaf cells – details only visible if I slice leaves into sections no wider than a human hair and magnify them 100 times beneath a compound scope. Given that each *Crossidium* plant bears only a few leaves, naming this moss risks destroying any specimen I might collect.

I stand up and the *huchar, huchar* of California quail floats up from below. Today, I am alone; last week, three of us – me, Marc, and our friend, Terry McIntosh, one of BC's dryland moss experts – surveyed further north. *Crossidium seriatum* is a North American species, but in Canada it's known only from 15 sites in the southern interior of BC. Within this restricted geography, it grows only on the fine-grained silts deposited in the bottom of post-glacial lakes and now largely eroded into steep-sided cliffs.

Above me, swallows swoop; out over the lake, an osprey soars. Maybe it's my solitude, or the overlap between *Crossidium*'s range and my childhood, but this morning's search feels caught in a tangle of home and history, of naming and losing. Thirty years ago, this rolling mix of fields and ponderosa pine forests, of open grasslands and blue lakes, were the contours I longed for when my mother's spontaneous marriage moved us south to Montana. If I'd gone even further afield – to the deciduous forests of Vermont – to learn botany's nomenclature, the universality of its grammar helped me return to BC – first temporarily as a doctoral student, and then permanently with a faculty position.

What turns country into home? It's an old question, but one especially difficult to answer for a rare moss. What are the limits of country when your body is permanently anchored in one place by thread-like rhizoids? How does time beat when you can curl up and go dormant, not just for a season but for multiple years, even decades, before the gift of water plumps you awake again? What is touch for those who live with little or no epidermis – absorbing nutrients and water across their entire bodies? All plants, but perhaps bryophytes most of all, test our understanding of body and being, home and place.

Still, this type of imagining has deep roots in botany. Some say it was first illustrated in 1804, in the diagram Alexander von Humboldt published to summarize his team's attempt to climb Chimborazo, Ecuador's highest mountain.[1] In this tableau, 13 columns, listing the environmental variables measured en route, bound the profile of Chimborazo. I've never been to Ecuador, never even been south of the equator, but I love this profile for the abbreviated Latin names running amok over the mountain's surface. Looking, I am forced to move between plant name and variables – catapulting me into the same type of conversation I imagine Humboldt's team having had on the mountain. I become a co-traveller, brushing up against the plants, feeling the prick of their spines and the soft brush of leaves, smelling their fragrance, sensing the pressure and temperature changes that always accompany an increase in elevation.

Several hours later, my hands and clothes are covered in a fine silt, and my lower back complains. It's time for a break. I sit on a mostly horizontal ledge and look out. From a distance, the contours of this valley are the same as I remember, but I know that, on the plateau immediately above me, vineyards and orchards have left little space for what was once native grassland enlivened with crust.

Botany may have let me return to BC, but at what cost? In North America, botany is part of what burst, in the words of Indigenous poet and scholar Jeannette Armstrong, "out of the belly of Christopher's ship, running in all directions, pulling furs off animals, shooting buffalo, shooting each other left and right."[2] Historians describe my professional ancestors as "agents of empire."[3] In 1833, when the Scottish plant collector David Douglas drew maps of this valley, this lake, his work served the interests of the Horticulture Society of London. And even though some collectors like Alexander von Humboldt protested the use of science to justify imperialism, historians explain that inventory sciences like botany clearly supported the expansion of Canada and the United States from one side of the continent to the other.[4]

I don't disagree, but I worry that perceiving botany *only* as a tool of empire might underestimate the power of its fieldwork. The most thoughtful collector I've ever known was my professor, Dr. Wilf Schofield. Wilf was already retired when I met him. But in the late 1990s, his office – houseplants clambering up windowsills, CBC Radio 2 playing, teakettle boiling – provided good habitat for me within the University of British Columbia's botany department. One of the foremost bryologists in North America, Wilf, along with his students, spent years mapping the homes of this continent's bryophytes. I knew none of this when I first met Wilf. He was merely the elderly gentleman my research supervisor sent me to see when I suggested bryophytes as a model system for my dissertation. Certainly, I had no idea that, for the next six years, Wilf would be the taxonomic cornerstone of my PhD research.

I'd arrive in Wilf's office with a shoebox of mosses I couldn't name, and in between guiding my identifications, Wilf would tell stories. Some were about his daughters, or the rare book he'd found, half-price, in a local bookstore. Others arose from his long history with plants. During his 47 years at UBC, Wilf spent most summers in the field. Sometimes, as we sat at his microscopes, Wilf, peering down at a moss, would ask, "Tell me again where you got this?" and then turn to grab the sheaf of maps he kept in his office – each one plotting the distribution of a bryophyte species within the province. I never knew who was more excited – Wilf or me – when one of my collections resulted in a new dot on a map.

Midway through my research, I needed an estimate of bryophyte species' fidelity to their habitats. I gleaned as much as I could from Wilf's publications, and then sat with him and went through my list of species, one by one. He knew them all – not just their names but the shape and textures of their individual homes. In more than 100 species, I didn't stump him once. Maybe it was then I recognized how thoroughly mosses had shaped Wilf's mind.

BOTANY MAY BE A TOOL OF EMPIRE, but today it gave me reason to go outside. Reason to spend this morning in May visiting the valley I once shared with my mother.

My mother. May. Mother's Day!

For a moment, I panic. Then I remember the cell phone in my pack. Alexander von Humboldt had to wait for his mother to die before he could go plant hunting; more than 200 years later, my mom sounds thrilled to be called mid-hunt. As we contrast her memories to what I can see, I use my binoculars to track the flight of an osprey. Something white flashes beneath the green canopy of cottonwood leaves immediately below me. An osprey on the ground? A gull? No, it's not feathers, it's skin...old skin...it's an old man's bum, lifting, about to roll over...

"Holy shit," I gasp, dropping my binoculars, standing up.

"Mom," I blurt out, "I gotta go." My mom's concerned, thinking maybe a bear, but I'm talking over her, ending the call before she can say goodbye.

I need to move. Now.

It's not the nudity that's distressing; it's what the eyes belonging to the elderly genitalia I just saw in excruciating detail will think if they see me peering down from above. Plant hunter or peeping tom? The difference might be hard to explain.

But I'm only a few steps along the trail before I'm wishing I could share this story with Wilf. This would be one to rival the many he told me.

But I can't.

I've known this moment would come. When the phone rang several years ago, I drove west over the mountains in a rush, afraid I would be too late. I was not the only one. During his final days in hospice, students from across North America came to stand beside Wilf's bedside. It was all too easy, with him already gaunt and skeletal, to wonder how our botany would survive his absence.

I'M ON MY HANDS AND KNEES AGAIN, feeling pressured, the sun now masked behind gathering clouds. This trail seems like perfect habitat, but no sign of *Crossidium seriatum*. A little further, I think. Nope...nope...nope. I've only ever found this moss once on my own. Why did I agree to this survey? Surely, I have better things to do than crawl on hands and knees above a beach occupied by naked men. Nope...nope...nope...frustration wells. Nope...wait. There? I grab for my squirt bottle, exploding artificial rain on tiny architecture.

I reach out and pluck the dripping scrap of crust. Beneath my hand lens, the crust swells. Is that it? I look up, world zooming out, twist the crust around to get a different angle and look down, zooming back in. My world resolves into microtopography, complete with gullies and peaks, canopied to one side by taller mosses. I scan right, sliding over an exposed plain of watery silt, and then, there, half-buried in a gully, a silvery whip of white. *Crossidium seriatum.* I look up, twist, look down. Yup. Glee burbles up, but I abruptly tamp it down. Not yet.

I slide the tiny sliver of crust into a brown paper bag. With a thick black pen, I write *Crosser(?), Naramata Rd, 3 Mi Pt*, fold the bag in three, and place it in the top of my pack. I GPS my location, write a habitat description in my field book, noting the recreational trail within the cliffs. Standing up, I search the rest of the silt face. Yup, there, and maybe there too. I won't know for sure until I can get it under a microscope. But I think I've got it.

I can go. As I pick up my backpack, triumph wells up, spilling over. I know better than to tempt the gods. But it's hard. I'm awash in the same joy I saw so often in Wilf's eyes.

Specimen stowed in my backpack, I walk back the way I came, thinking about the relationship between collectors and the land from which they collect. It was on a spring day not unlike today that David Douglas, en route from Fort Vancouver to Fort Simpson, botanized along the long ribbon of this lake. He, like me, would have used Linnaean nomenclature – a grammar foreign to this valley – to name the specimens he collected. By privileging communication across time and space, the grammar he used left little space for the local or the Indigenous. David Douglas and Alexander von Humboldt botanized opposite ends of the

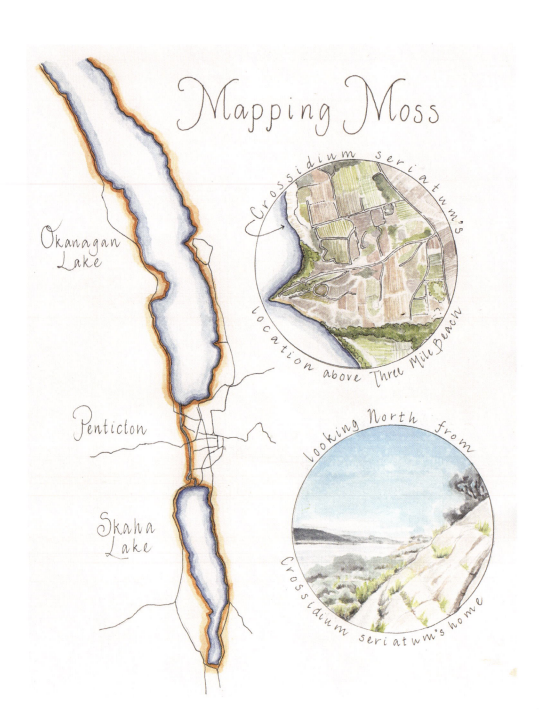

Mapping Moss

Okanagan Lake

Penticton

Skaha Lake

Crossidium seriatuum's location above Three Mile Beach

looking North from Crossidium seriatum's home

Illustrated Map: Mapping Moss

so-called New World, nearly 30 years apart. But both had their mosses identified by William J. Hooker, the eventual director of the Kew Botanical Gardens, and a man who never set foot in this valley. More than 100 years later, Wilf used the same language to share the specimens he collected with museums around the world.

I've long understood the benefits of Linnaean taxonomy; I've only begun to understand its limitations. It's not that the specimen I found doesn't matter. It does. Labelled and deposited in the same herbarium that holds Wilf's collection, this specimen will help delineate the home of an endangered species. But, today, my worry about the very real horror of Empire that haunts botany's nomenclature is matched by my worry about botany's marginalization within our society.

In the 21st century, the general public's *experience* with native plants is nearly as endangered as *Crossidium seriatum*. A hundred years ago, when plant collecting was considered recreation, in this valley there would have been men, women too, who would have recognized my collecting gear; some who regularly sent their own specimens to provincial botanist John Davidson.[5] And before European settlement? I don't know if Jeannette Armstrong's ancestors would have called *Crossidium seriatum* by name, but I don't doubt that their everyday intimacy with plants far outstripped mine.

In today's globalized world, when avocadoes and lettuce from Mexico comprise more of our daily menu than native berries, does it matter if we know what counts as being home for a moss, rare or not? I think it does. Imagining another species' homes is like climbing a mountain; the real goal is not the summit but the map we make en route. For it is within the moments of uncertainty, of figuring – here, not there; on this slope, not that one – that we taste the world's silt and sand and blood, that we allow the world to imprint upon us. On this, I am sure: few plant collectors return unchanged.

I'm nearly at my car. I dread the next few hours – the long, slow drive north through a valley congested with townhouses and shopping malls. As a species, we humans are many things, but rare we are not. Maybe, I think, if we let more species *imprint* upon us, we might moderate the imprint we make on others.

I dig out the specimen I've collected from my backpack. In my hand, it feels less like a specimen and more like a thread of hope. Just for a moment, I try to imagine a map of home that could name without claiming, that could reconcile data with experience, Indigenous with settler. Like all maps, it'd be flawed – an incomplete, inadequate and imperfect rendering – but I think Wilf would lean forward to look.

8. Say the Names

TWO COASTLINES, two maritime landscapes. Neither is where I live, but one is home, and one is not. In less than a week, the juxtaposition of work and holiday has me flying from the tortuous bays and inlets of Nova Scotia to the isolated fields of Washington state's San Juan Islands.

On one Tuesday, I am on my hands and knees, amid a tour of botanists, exclaiming over the chartreuse, red-veined, vase-like leaves of carnivorous pitcher plants in the coastal barrens adjacent to Polly's Cove. I am in good company. The father of taxonomy, Carolus Linnaeus, is said to have described botanists as "those much given to exclamations of wonder."[1]

In the odd misty light of the coastal barrens, excitement bounces off stunted vegetation.

"Here's the huckleberry, *Vaccinium macrocarpon*."

"Look, one of the pitcher plants is in flower. Oh, here's a whole patch of them in flower."

My hands and knees are wet from kneeling in the soggy ground of the barrens. I look up around me. A gray-haired botanist whose name tag is obscured by the immense camera he's carrying, and who I don't know but instinctively like because he's included his 11-year-old daughter in this adventure, is parsing the odd arrangement of this species' floral parts. "If these are the sepals, and this modified extension is the style, where are the petals?"

Nobody's shoes remain dry.

The next Tuesday, Marc and I, amid a crowd of spandex-clad bicyclists, wheel our bikes off the ferryboat, the MV *Elwha*, onto the rocky shore of Lopez Island, one of the San Juan Islands crowded into the Salish Sea. A young ferry worker, her orange safety vest individualized, island-style, with bead and lace trimming, legs planted firmly in the centre of the narrow asphalt road, repeats over and over again, "Bicyclists, please wait in the parking lot for the cars to unload. Thank you. Bicyclists, please wait. Thank you."

Marc and I have neither spandex nor the sleek outlines of high-tech touring bikes, so we are happy to wait not just for the cars but for the other bicyclists to speed ahead of us. This is our first holiday alone since Maggie was born nearly nine years ago. As we climb the first hill out of the ferry landing, we cycle side by side and fall into the rhythm of the language that first brought us together.

"Is that *Lathyrus* there in the ditch?"

"Yeah, but I don't know if it's *nevadensis* or *japonicus*."

"The *Holodiscus* is in full bloom – I haven't seen it like this in years."

Without child, without jobs, we are just us, again.

Two maritime landscapes, two coastlines. One home, one not. One is of my country, one is not. Surprisingly, home and country do not correspond.

The word "flora" describes both the collection of plants found in a region and the technical manual that lists and describes these plants. Many technical floras contain maps of individual species' distributions. Flip through these pages and it's clear that the ranges of many species overlap. Others are less congruent. There are no hard and fast rules about the borders of a flora. Some manuals describe all the plants found within the limits of a state or a province. Others describe a group of plants that the authors believe to be a cohesive regional flora, regardless of political boundaries. Botanically, I grew up with the one-volume version of the *Flora of the Pacific Northwest*. My battered copy of this flora describes the plants in an irregular geographic region centred over the state of Washington, extending north, the authors say, to include "an indefinite fringe of British Columbia," east to the mountains of Montana and south to the northern half of Oregon.[2] Published by two grand old men of North American botany, Leo Hitchcock and Arthur Cronquist, in 1973 when I was just 7 years old, this book has long been my authority for the names, common or scientific, of flowering plants in this area.

Say the names. Names elicit knowing. Names breathe a landscape into being. Names define home.

I have lived all my life in the borderlands of the 49th parallel, flipping from one side to the other with time and necessity. Throughout my childhood, my mother's marriages were associated with name changes not just for her but also for my siblings and me – obscuring any trajectory my biological father might have

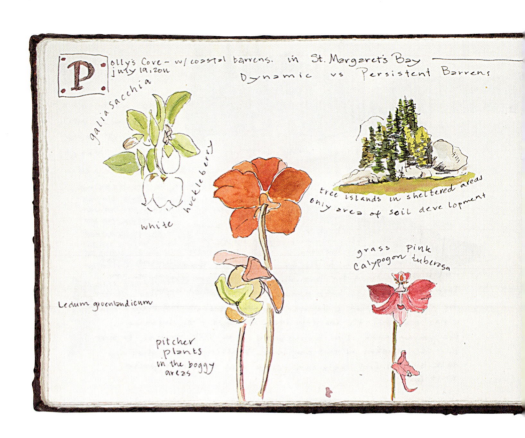

P olly's Cove - w/ coastal barrens. in St. Margaret's Bay
july 19, 2011   Dynamic vs Persistent Barrens

gallia Sacchia

huckleberry

white

tree islands in sheltered areas
only area of soil development

grass Pink
Calypogon tuberosa

Ledum groenlandicum

pitcher
plants
in the boggy
areas

spencer spit

Vol. 19: Polly's Cove, and Vol. 19a: Spencer Spit

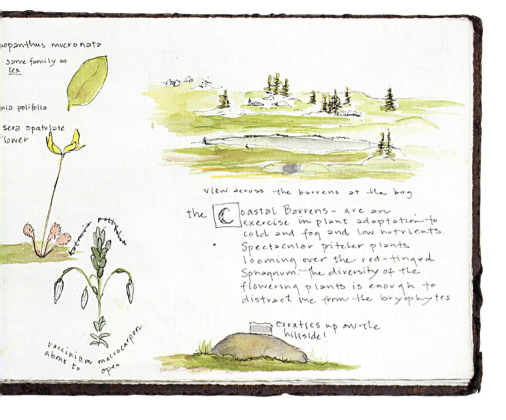

opanthus mucronata

same family as
Ilex

nia polifolia

sera spatulate
lower

View across the barrens at the bog

the [C]oastal Barrens - are an
exercise in plant adaptation to
cold and fog and low nutrients.
• Spectacular pitcher plants
looming over the red-tinged
Sphagnum. The diversity of the
flowering plants is enough to
distract me from the bryophytes

erratics up on the
hillside!

vaccinium macrocarpon
about to
open

2.7,   2 o4

sought. My faded school report cards track the changes in nomenclature: *Baldwin* in kindergarten, *Reynolds* in Grades 1–5, and *McInturf* in Grades 6–11. Only in my final year of high school did the necessity of completing university application forms allow me to reclaim my first and only legal surname.

Far from home, I learned the grammar of scientific naming, first codified by Carolus Linnaeus in 1753. All species, microscopic or macroscopic, plant or animal, are given a two-part name, termed a *Latin binomial*. Today, all botanists, regardless of their native language, know Douglas fir trees as *Pseudotsuga menziesii*. There is a set of rules, deliberated every six years by an international gathering of botanists, governing the naming of plants. While some names simply describe, others commemorate. *Luzula spicata* depicts the spike-like arrangement of the flowers of one grass-like plant, while *Luzula hitchcockii* remembers Leo Hitchcock. Each species has one correct Latin name, but sometimes *correct* changes with new knowledge or new arguments. I'm not good with name changes. In Nova Scotia, I discovered that one group of eastern huckleberries that I'd long understood as *Vaccinium* are now called *Gaylussacia*. I was happy to retreat west, where *Gaylussacia* is absent from my flora.

Say the names. Names have history and order. Names identify and also obscure. Names link past with present.

My legal surname, Baldwin, is of English origin, from the Old English, *Bealdwin*, or the Old German, *Baldavin*, and means *bold friend*. I wish the small, fatherless girl I once was had known that. Even though I don't know my father, I've become fond of the Baldwin name, cleaving to its two distinct syllables.

Marc and I met working as itinerant botanists in western Montana. I first felt my husband's hands when we were using fingers and palms to physically describe the difference between *compressed* and *obcompressed* mustard pods. That first summer, we were sent to survey different valleys, different plants, and wrote to one another from the isolated shelter of tents. On our break days in town, attraction flourished over the smell of a herbarium collection. Marc came with me when I started my PhD in BC, and I returned with him to Montana when he landed a full-time ecologist job – crossing and recrossing that arbitrary line of country and mind. Our daughter, Maggie, was born just ten miles east of the Continental Divide in the historic gold-rush capital of Montana. Maggie's legal name carries the full weight of our affiliations. *Rowan Margaret Baldwin Jones* pays homage to strong trees, strong mothers, and fathers known and not.

Marc asked me to marry him not with a ring but with a bracelet woven of flowers from my favourite plant family – the Ericaceae or Heath. This group of plants includes a collection of delicate, urn-shaped wildflowers, most of which develop into fruits like sweet huckleberries or crisp cranberries.

I've told this story at dinner parties. Marc will remain mostly silent, adding only, "I was a starving botanist. What else did I have to offer?"

A man who knew the language of plants. A man who didn't need me to change my name. The offering was immense, my decision immediate.

Some people bring pets or children to a relationship. Marc and I brought floras. Today, our bookshelves strain under the weight of technical manuals. We each have our own copy of Hitchcock and Cronquist – bought before sharing was imagined. Together we invested in the eight-volume *Illustrated Flora of British Columbia*. While the weight of our shared flora has coalesced upon a common region, our individual floras still differ. I ask Marc to help identify wetland sedges, and he leaves all the mosses to me.

For botanists, knowing floras can translate into paid work. Time spent in Vermont and BC increased my prospects in both places. When I was offered a permanent faculty position in Kamloops, Marc supported the move, relieved he wouldn't be asked to locate east of the Missouri. A change of country was preferable to a change of flora.

Say the names. The names of plants – the scientific or common name, the Western or Indigenous name – are a poetry that links together a history of people looking at plants. Plants become known when there is reason to know them.

During our first day in the San Juan Islands, the melody of a known botany – lupine, reed canary grass, Scotch broom, salal, sword fern, bracken, Douglas fir, grand fir, *Spiraea*, *Holodiscus*, self-heal – played through me as we cycled. Not all were native – some were even noxious – but all were easily recognized. If I didn't know it, Marc did. When I walked last Tuesday through the coastal barrens and deciduous forests of Nova Scotia, there were only faint echoes of familiarity. I knew the huckleberries and the pitcher plants from similar wetlands in Vermont. But the grand chorus of the dominants was missing. I stumbled on the difference between ironwood and beech, between black and red crowberry.

The distribution of plants is oblivious to the arbitrary lines of political borders. I imagine my home addresses charted as individual dots on the map of North America – a personal equivalent to a species distribution map. My history has the greatest density and pull in the intermontane west. With a few vagrant dots to the east, these locations become increasingly tangled, weighted with green things and with Marc, and then with Maggie, in a region that has no name, yet largely corresponds to the geographic area encompassed by the *Flora of the Pacific Northwest*. After a week on the eastern margin of our continent, arriving in Washington's San Juan Islands – even though it meant crossing that infernal 49th parallel – was returning to a land of the identifiable and the shared. This western archipelago is not where I live, but it shares enough botany with interior BC to have the smell of home. My home range overlaps that of no individual species precisely, but there is an assemblage of species that marks my terrain, a known flora that shifts country toward home.

PART 2

Collecting Kin

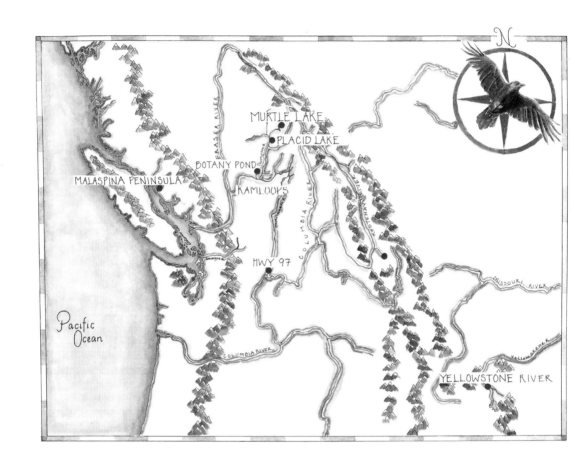

Location Map for Collecting Kin

9. When Mountains Move

 WE PADDLE ACROSS the big blue water of Murtle Lake in BC's Wells Gray Provincial Park, just the two of us: me and my friend Libby. Two days ago, we abandoned our respective homes – mine, four hours south, Libby's nearly nine hours distant in the lowlands of Washington state.

Driving north, kayaks piled atop Libby's car, gear stuffed in the rear, I'd known how my absence would burden Marc with more than his share of child care, dog walking and housecleaning. Yet the moment my body felt the first rhythm of paddling – dip, push forward with opposite arm, dip, push forward – I could only be glad. This is my first wilderness trip since Marc and I decided, kayaking on BC's coast, to have a child. Maggie will turn 11 years old next month. For nearly 12 years, the burden of pregnancy, the full-throttle of early childhood care, and then the demands of a new job have tied me to the front country.

Off my bow, Libby paddles steadily. In front of me, mountains run from lake to sky, their reflections pooling in calm water. With no obvious human imprint, the view feels like a salve I'd nearly forgotten to need. I never meant to be gone for so long.

If, before Maggie, my life revolved around wild places like this one, 12 years has been long enough for me to forget that success in the backcountry depends as much on what you leave behind as what you bring. I know: the evidence is leaking in my forward hatch, oozing up against freeze-dried rice and prepackaged curries. Unable to imagine a week without fresh vegetables, I'd plotted Tupperware dimensions against the volume of my kayak's hatches, counting out peas and carrots and avocados and red peppers. But vegetables belong in the garden, not here. Two days in, and my vegetables have already collapsed into a mess of crushed skin and bruised flesh.

Here, where the landscape opens, paddle stroke by paddle stroke, where fingers of glassy lava slide into shadows of blue, where cedars live for centuries without the touch of human hand, even the word "vegetable" sounds out of place. The shape, colour and taste of all vegetables reflect their allegiance to a single species – us. A "botany of desire," as Michael Pollan puts it.[1] There's no doubt that by hitching their stars to ours, the plants whose roots, stems or fruits fill our kitchens have expanded far beyond their native range. But at what cost?

Today, geographers argue that the path to the Anthropocene was paved, at least in part, with vegetables. That is, as Europeans first transported, and then cultivated, plants like potatoes, wheat and corn across the world, they laid the caloric foundation that subsequently allowed our species to grow into a planetary-level influence.[2] Globally, our impact now rivals that of the Pleistocene's glaciers, but we depend upon a botany divorced from place, one that has transformed plants from living beings into inanimate commodities that can be bought and consumed with little knowledge of the soil in which they were grown.

In the Anthropocene, even our botanical imagination has been globalized. Hundreds of thousands of millennials follow "plantfluencers" via Instagram and Pinterest,[3] but few who appear in my botany labs can identify the native species found in the sagebrush steppe that surrounds Kamloops. Even fewer know of the rich, wet interior rainforest that sits in the headwaters of our watershed.

To be honest, I'm conflicted. As a botanist, I'm delighted by any trend that counteracts the plant blindness that has worried my discipline for decades. But, as an ecologist committed to the botany of BC, I'm concerned when we appear to care more about plants selected by industrial supply chains than by the climate outside. If naturalist Robert Michal Pyle was right – that any effort to conserve natural ecosystems or species must begin with rebuilding our experience *with* them[4] – then we need to rebuild experience with *this* flora. The question, of course, is how?

Few, I think, would argue with the importance of direct contact. Here, in Murtle Lake, I've been revelling in a botany that is clearly rooted in place. Western red cedar, black cottonwood, Douglas fir and western hemlock: all species whose history with this landscape, this part of North America, far exceeds my own. But rebuilding experience with a flora demands more than just listing its species. The word "experience" comes from the Latin *experientia*, meaning "a trial, proof, or experiment." The work that has most tested me, of course, is the mix of art and science contained within the cover of my field journals.

I know field journals like mine were an important tool in the European project to name, claim and transport the world's botany, as their entries made the lives of rooted plants more visible and definitely more portable. Historian Mary Louise Pratt defines *contact zones* as "the spaces of colonial encounters, the space in which peoples geographically and historically separate come into contact with each other and establish ongoing relations."[5] Historically, such contact zones resulted in conflict, coercion and radical inequity between the colonizers and the colonized. Yet contact zones are never one-sided. Historians and geographers tell us that European illustrated travel accounts, like any act of mapping, were "creative, sometimes anxious moments in coming to the knowledge of the world,"[6] that the writing and sketches collected in these journals were never completely objective. In creating these travel accounts, many Europeans appeared to have become entangled in a series of difficult negotiations between their assumptions and the *experience* of travelling the field.[7] Certainly, since returning to Kamloops, no tool has been more important in storying a place, maybe even a home, for me.

Last year, in writing the application to fund this trip into Murtle Lake, I explained that I couldn't rebuild *experience* with plants with a sampling quadrat, but I might with an illustrated field journal. Given the role that field journals had played in the European colonial practice that helped separate us from them, ecology from wealth, culture from nature, what would be more fitting, I argued, than to co-opt their form to help rebuild our experience of plants in place?

It worked. Or at least my proposal succeeded. That's the reason that, on a Tuesday in mid-September, I can be here with Libby, one of my first field journal mentors, rather than in a lecture hall. The reason art supplies fill an entire dry bag in my kayak. The reason I've come to this lake.

Murtle Lake. For nearly a decade, I've heard its praises: the largest motor-free lake in all of North America; a lake bounded not by roads but by an intact, inland rainforest, dripping with moss, resplendent with big trees. A lake accessed only via a two-and-a-half-kilometre portage; a lake whose waters run free for 200 kilometres before flowing through the heart of Kamloops. In looking for an experiential lodestar, no ecosystem seemed more fitting to journal than the wild, wet botany of Murtle Lake.

But, two days in, all I've got is an odd aphorism that pops into my head as loons call off my bow and dragonflies zip across clear water: *"It's important to go where vegetables sorrow."*

Yet how is my impulse to carry vegetables into Murtle Lake any different than the human desires that carried European crop species to North America? The bruised vegetables in my kayak are a rebuke: If I've insisted on carting vegetables into the wild, do I really think I'll be able to gather stories to compete with the slick botany of an Instagram feed? What the hell am I doing?

I have no choice. In less than a month, I'm scheduled to give a reading; in three months, my art show at a local gallery will open.

The next day failure dogs me, dawn to dusk.

Libby and I had planned to spend the day drawing along the trail into McDougall Falls, the first of six waterfalls found below the lake. But when Darryl, the park keeper, visited us in the morning, and tells us that a pack of wolves had been spotted close to File Creek, we instantly reorganized our day, spending all morning securing a tent pad in the campsite closest to where the wolves were seen. And then the trail into McDougall Falls was both further and slipperier than we'd imagined. The result: a nearly blank page in my field journal. Other than an abbreviated plant list, the only worthwhile botany of the day is the collection of king boletes Libby gathered along the trail as we were rushing back to our kayaks. Not even the abrupt howl of wolves, one exuberant declaration just at sundown, breaks through my growing disquiet.

On some botanical expeditions, imminent failure engenders heroic acts. For me, it elicits the absurd gesture of opening my field journal when it's nearly too dark to see.

As stars spill from the sky, I use a pencil in my outstretched hand to measure the distance between the dark mass of Central Mountain and the stars nearby. Shifting from horizon to paper, I translate my measurements into guide marks. Draw the mountain's profile. Recheck the measurements. Something's off.

Measure again. Mark. Draw. Error.

Frustration builds with each wrong line, until abruptly the understanding I have of my world slips, falters, and then, for an exquisite moment, fails.

In front of me, a mountain moves.

I've made no errors. The lines I've drawn are not mistakes but evidence of the mountain's progression. First toward, and then past stars. I continue moving my pencil – less for the marks on the page than for the sheer novelty of watching a horizon in motion.

Of course, the mountain doesn't really move; the earth rotates. But no abstract diagram of Earth's rotation has ever carried the shock of these past few moments. Science may live with a Copernican understanding of our solar system spinning in space, yet most of us, I think, live with the experience, the day-to-day expectation, of a stationary world. But here, in the deep dark of Murtle Lake, there's no escaping the evidence.

Mountains move. Not just against the stars. Up-thrusting with the collision of continents, subsiding with erosion, carried past latitude or longitude on the backs of shifting plates, their form, at any one moment, a temporary hesitation of elements on the move – cycling from rock to water, to cloud, to mesophyll, to muscle, and then back again. The mountain, and, by association, the whole world, is alive.

Now. For me. For us. Later, before we crawl into our sleeping bags, Libby and I decide we've been travelling too fast. Tomorrow we'll stay in place. Early – very early – the next morning, birds sing me awake, and I open the tent fly to peer across the bay. And there they are: five wolves, black as ravens, wrestling in the slanted sun. They don't stay long, disappearing before Libby can wipe the condensation from her binoculars. But it's enough.

Wolves playing. Libby and I drawing. The living shore of Murtle Lake fills my field journal. Sunlight breaks jagged over distant conifers; glossy black seeds spill from violet capsules; paintbrush burns red against golden grass; wings of sandhill crane beat against the air immediately above me. When Libby finds a small wildflower whose name I don't know, I fall into its complicated morphology, drawing it in detail.

I am so immersed that it is nearly noon before I register the events taking place offshore. Slap...slap...slap. With my binoculars, I scan the lake.

Osprey. One after another they plunge, talons outstretched, thwacking the water's surface with their full weight. Milliseconds later, they're shuddering upward into flight, rainbow trout flailing in their grip. In the hour I watch, no osprey is denied.

Later, when Libby and I paddle toward the wolves' sandspit, an osprey flies past me, talons wrapped around trout. I turn in my seat to follow it back to its nest. Pulled close through binoculars, a hooked beak tugs at flesh. An implacable, yet

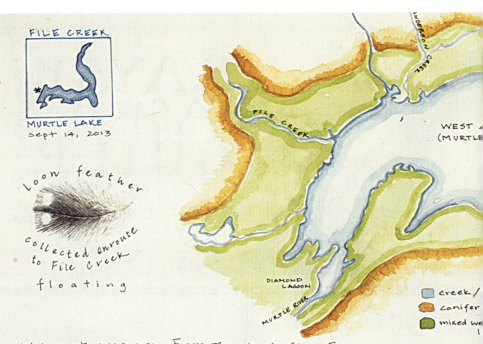

FILE CREEK

MURTLE LAKE
sept 14, 2013

loon feather

collected enroute
to File Creek

floating

ANDERSON CREEK

FILE CREEK

WEST A
(MURTLE

DIAMOND
LAGOON

MURTLE RIVER

creek /
conifer
mixed wo

Wolves at 6:00 am. From the tent fly. Five wolves coloured
black prancing and wrestling on the sand in front of Anderson
creek mouth. Once again, the feeling that I have been
gifted with something wild and precious rises up, rises up.
The world is alive and I am caught within its own wonderful
It is a remembrance of another time, when we hadn't
domesticated ourselves, when the glory and the pain and the
deep splendor of a breathing world filled our days.

As I write, the sound of large wings fill the air, I look
up to see two large birds, sandhill cranes flying west. No ca
no trumpeting sound, just feathers brushing through
blue sky before they are lost beyond the canopy of spruce and fir.
I sit at the south margin of File Creek's mouth. Fingers of basa
reaching out to lake water; a gray speckled moth slides into the
crack of a dead spruce ~ instantly camoflauged kinglets chitter high-
pitched above me. This richness nearly deafens all my senses.

Wells Gray Vol. 4: File Creek

view across the mouth of File Creek

Lobelia kalmii

yellow

white

purple/blue

linear leaves

viola capsules

tiny black seeds

open, seeds lost

green, still forming

parnassia fimbriata

graceful, choreography. Twist, tear, twist. First, silver skin, then blood-red muscle and finally slick viscera disappear.

Beneath me, my kayak scrapes against sand. I'm still well offshore, but when I step out, the calf-deep water explains the ospreys' success. Kokanee – landlocked salmon – spawn in File Creek this time of year. Their congregating masses, and the eggs the females lay, attract predators, including rainbow trout. But the same creek that provides the kokanee with spawning habitat has filled this end of the lake with sediment, thinning the watery boundary between trout and osprey.

In the soft sand of the spit, Libby and I find tracks of eagle, raven, sandpiper – and wolf. It's easy, sketching the outline of a paw, to imagine the muscle and sinew that pressed earthward, the gait that swallowed distance. One set of tracks points like a compass arrow, a glyph of wolf intent, that's impossible not to follow. Off the beach, an old stream channel narrows into thickets of willow and alder.

Maybe it's the powdered bone in the wolf scat we find, maybe it's the alder walling us in, but suddenly I wonder about the wisdom of walking *toward* wolves. A moment later, there's a small *whoof* – not the howl of the night before, but definitely a warning. Doubt vanishes in a cold clench of fear, and without words Libby and I turn back.

In the kayaks once again, with a clear sightline between us and the wolves' napping spot, I open my field journal and drift, floating in the polar attraction of the world. Mergansers, grounded by moult, foot-paddle their way across a bay brimming with life, through water lit from below by a sinking sun.

Paddling in, I'd seen this end of the lake as little more than something sitting between me and a tent site I wanted. Only 24 hours later, it tangles with complication. The water under my kayak is a killing ground for rainbow trout, a lardor for osprey. The creeks running from mountain to lake are birthing chambers for kokanee, an all-you-can-eat buffet for wolves. The wet forest we hurried through yesterday is sewn together with the mycelia of boletes, and fertilized by the rotting carcasses of kokanee dragged from stream by wolf or bear. *Where vegetables sorrow* is, of course, also *where predators prey*. Not just a place, but a way of being that understands that all flesh – botanical, fungal or animal – will be, and should be, the stuff of someone else's feces.

Picking up my paddle, I turn back toward the campsite.

Yesterday, I doubted my project when I focused on the incongruity of having brought carrots, a species so entwined with humans that I've never wondered about its native range, into an ecosystem where so many species survive in spite of, rather than because of, human intent. But the wild abundance of Murtle Lake's top predators is, like my carrots, a reflection of both biology's potential and human fiddling. Today, Murtle Lake's wolf and loon and osprey feed upon the red bodies of kokanee and trout. Yet, for most of this ecosystem's history, no fish swam here. The waterfalls downstream are too tall for kokanee and trout to jump, the lake too isolated for osprey or eagles to have accidentally released their wriggling bodies into its waters. Humans *preyed* this lake into being when, nearly a century ago, government biologists dropped first trout and then kokanee from the belly of low-flying aircraft.[8]

Does it matter if the trout being hunted by the osprey overhead belong as much to the Anthropocene as the rising levels of carbon dioxide in the air? Not to the osprey. When sand-shallowed water offers up silver-sided fish, they don't hesitate. They drop. Plunging through air and water to seize their prey.

In coming to Murtle Lake, I was grasping for a place apart, a botany unsullied by human influence. But any search for separation, seductive though it may be, will fail. Not only is this ecosystem part of the same traditional territory – Secwepemcúlecw – that envelops my home in Kamloops, but its plants have been part of the Secwépemc People's story for millennia. If descendants of settlers like me, who know fewer stories about its plants, now also inhabit this territory, we are no less co-dependent. Plants and peoples: our futures depend upon one another. Last century, fishery biologists did not hesitate to fiddle with the top end of Murtle Lake's food chain; today, we need to learn to care for the green bottom of the world – both here and across the planet.

I wasn't wrong to come to Murtle Lake. I wasn't wrong to bring my field journal, or even my carrots. Places where vegetables sorrow, I now understand, are where the quicksilver flashes of uncertainty, of cognitive dissonance, germinate ecosystems of altered thought. The lesson of Murtle Lake lies not just in one taxon, or even in its isolation, but in its extraordinary, beautiful contradictions: how human tampering can make food webs more, not less, diverse; how failed lines on a page animate rock into motion; how *all* plants, in eating the sun, root the earth. I think back to my drawing in the dark, to my worry about the vegetables in my kayak. Exploring uncertainty – moments I first perceive as failure – will be, I realize, the surest path past the obstinate divisions of the Anthropocene: us versus them, Instagram feed versus wilderness experience, carrot versus cedar.

Eventually, I will understand this trip to Murtle Lake, not as a beginning but as a middle. In many ways, the work began the moment, years ago, when I first questioned my understanding of plants and place, home and country. In the years to come, I will wonder if my route to this work was defined by precise moments – say, like that of rainbow trout splatting into blue water – or if it was as tangled as the mutations shaping the curved beak of the osprey flying overhead.

Yet, as my kayak noses in beside Libby's, I can find little of the doubt that rose yesterday with the sorrowing of my vegetables. It's not that I underestimate the risks; I'm betting at least part of my academic career on my ability to navigate the wildland of art galleries and public readings. Certainly, I have no way to imagine how, in moving forward, the stories of plants in place will test my understanding of my family and kin. But, stepping out of my kayak and reviewing which of my vegetables may be salvageable for tonight's dinner, the risk feels an important part of this project. Etymologically, *peril* shares the same root as *experience* and *experiment*. No story gets told worry-free.

10. A Pattern
with Consequences

Men and words. My mother has rarely hesitated with either. For most of my childhood, she was serially monogamous and unafraid of saying what she thought.

I learned early to be wary of what could come out of her mouth. Fed up with me, or more likely with my brother, who looked too much like our biological father, my mother's voice would rise, spewing sentences – some lurid, others rude, many true.

"Black and blue" is one phrase that still lingers.

As in, "I'm going to beat you black and blue and rip your lungs out!"

Girlfriends who were sleeping over would turn toward me, anxiety clouding their eyes.

"Don't worry," I'd whisper, "it's just something she says."

IT'S 4:30 P.M. Late July. Four of us – me, 12-year-old Maggie, Marc, and our new dog, Freya – stand atop the exposed granite rock of Sarah Point at the tip of BC's Malaspina Peninsula. The gregarious young man in flip-flops and cargo shorts who ferried us here (but whose name I've lost) waves once and pushes his water

---

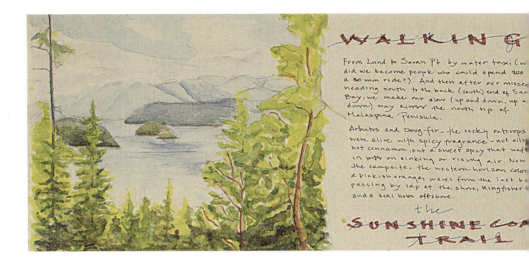

Vol. 30: Walking the Sunshine Coast Trail

taxi away. Offshore, blue islands hump up against the horizon. Distance shrinks the rumbling of the aluminum boat until a hot curtain of silence drops around us.

When I turn inland, a wooden sign announces the start of the Sunshine Coast Trail. We're not going far. Three kilometres to Feather Cove this afternoon, and then over the next two days, only 18 more to Malaspina Road, five minutes from the rambling farmhouse my mother now calls home. As we take our first steps on the trail – a trodden scar winding through crackly lichen and moss – I wonder how long it's been since I *walked* home to see my mother.

Now in her 70s, my mom lives with her husband, Bill, at the literal end of the road. From my home in BC's interior, the route to my mother's traverses two mountain passes, negotiates Vancouver's traffic, and then waits in line for not one, but two, ferries, before winding down the final stretch of Highway 101 into the small village of Lund.

After years of struggling to make this trip in December, most of my family now celebrates "Christmas in July" with Mom and Bill. Although summer visits risk fewer road closures and cancelled ferries, the weather inside my mother's house remains predictably turbulent. We all know the forecast: expect initial exuberance (seeded with abundant red wine), to be followed by increasing periods of annoyance (arising from the collision of overstimulation and physical

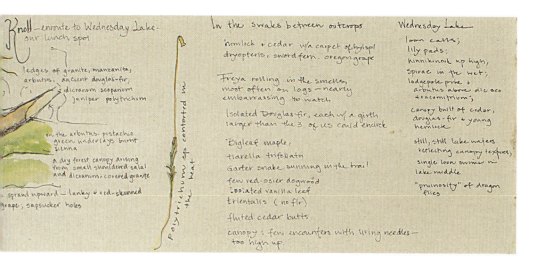

exhaustion), and ending with a high probability of tears (falling from one of us, never Mom) late in the visit.

Here's the dilemma: it's important to get out before the weather destabilizes, but if Marc, Maggie and I visit for less than a week, I hear about it for the rest of the year.

The compromise: we arrive with my extended family, but when everyone else leaves, the three of us go exploring – staying away for at least two days – before returning for a final night with Mom and Bill. Spending time in my mother's forest, even if not in her direct company, appears to count on the good daughter side of the balance sheet.

It works. Mostly. As I climb through the sun's glare, I wonder if Marc or Maggie noticed the lengthening pauses between my mother's sentences this morning. The wordless gaps that can still make me, a nearly 50-year-old woman, frantic to please. From one day to the next, no weather is predictable. Averaged over time, however, weather stacks into the climate capable of shaping individual lives, species, entire ecosystems.

Atop the first knoll, with a clear view north into Desolation Sound, I stop to wait for Marc and Maggie. Mixed in with the dry heat of Douglas fir, there's a spice – cinnamon mixed with pepper – I can't identify. I'm rubbing manzanita

A Pattern with Consequences

—

and arbutus leaves between my fingers, when I hear Marc and Maggie. Marc's recounting an old Irish tale to Maggie as they walk – an obvious, but apparently successful, attempt to distract Maggie's attention from a trail that is much steeper and hotter than we imagined. Not for the first time, I think how lucky I am to share the parenting of my child with this man.

It's a long-faded photo. Mom and her about-to-be second husband, Simon. Him in a blue satin shirt, trimmed with embroidery around the edges, her in a cream gown, again with embroidery. Simon's drinking from a glass bottle, lips wrinkled against the bite of vodka and grapefruit juice. I have little memory of this moment. My siblings and I were hiding, too embarrassed to watch our mother's hippie wedding in Vancouver's Queen Elizabeth Park. This would be the only one of my mother's four weddings I attended. She married my biological father before I was born, her third husband, Charlie, without telling us, and Bill during my final exam week at university. Thirty years later, my mother has yet to forgive that final absence.

IT'S DAY TWO. Perched on granite bones dipping down to the sea, the campsite at Feather Cove was long on sunset, short on fresh water. This morning, we had to hike two kilometres before we could fill our nearly empty water bottles. All of us, Maggie especially, have been coming to grips with the undulating topography of this forest. Down, up, down.

It hasn't helped that Marc and I – anticipating another hot afternoon – turfed Maggie out of the tent long before she was ready. But today the heat has built quickly. The last few hours have been more trudge than stroll; Marc and I trading who walks with our whiny kid, and who walks ahead with Freya. In the distance, I see Marc and Freya waiting for us at the base of the next hill. As I anticipate trading kid for dog, my mood lifts. But the moment we draw near, Maggie collapses on a log and refuses to move.

Hot irritation surges through me; each minute we delay now adds to those we will walk when it's even hotter.

I look to Marc for help, but the sight of his normally calm face tight with annoyance shocks me out of my own. Even Freya – typically buoyant – has already dug and curled into a shallow depression beside Maggie. Maybe, I think, Marc and I need to learn from kid and dog. Just a quick break, I suggest, slipping off my pack and sinking onto the moss-covered forest floor before digging out granola bars and water.

Around us, Douglas fir trunks climb skyward, collectively building a forest like few others. Sitting here, a full 50 degrees north of the equator, the presence of *forest* is not unexpected. Forests grow wherever enough rain falls to satisfy tree thirst. None of the trees near us are old, not even middle-aged – in size, only 60–70 centimetres in diameter at their base. But, left undisturbed, I know these trunks would swell outward for centuries, maybe even for more than a millennium. In North America, trees grow biggest – in both girth and height – in forests that face the Pacific.[1]

Rooted, trees link earth with sky; severed from their roots, trees build houses, sail oceans, bear totems of Clan and lineage. The bigger the tree, the greater our appetite for its wood. More than 100 years ago, loggers called the old-growth forest of this peninsula the "Jungles."[2] We've seen remnants of its original architecture – stumps shorn of height but still with girth too big to be encircled by Marc, Maggie and me, arms outstretched. Sitting here, I try to imagine the young men, many not born on this continent, who once stood on narrow springboards, rhythmically pulling misery whips, two-handled crosscut saws, across the grain of elders rooted in place.

Mom and I drive the Fortine-Wolf Crick Road in Lincoln County, Montana: her shifting gears in our puke-green Pinto like it's a race car, urging me to feel the straight line through the curves; me relishing her laughter. My mother's third husband, Charlie, and my siblings follow behind in his old International pickup. This would be one of the endless trips we made each autumn to get the firewood we burned all winter. Maybe I was 13. Certainly, it wasn't long after I had found Charlie's Penthouse magazines stashed in the carport. Growing up surrounded by hippies who wore nudity like a national costume, naked

bodies held little intrigue for me. But I had read all the Penthouse letters, disbelief battling with a curiosity I little understood.

On the road, Mom and I come sliding around a corner, and there is a rig pulled over, a snag bucked up in rounds beside it. Without warning, Mom screeches to a stop and opens her door. It's someone she knows. I get out of the car, still laughing from the ride, and she introduces me, "This is Lyn, my horny little redhead."

WHEN WE START WALKING AGAIN, the trail switchbacks up into the needle-filled canopy of the trees we just rested beneath. Needled or broad, evergreen or deciduous – there's more than one way to grow a canopy. I once heard leaves described as "the ephemeral skin of trees." I love this description, but as a botanist, I worry it misses the stark tension that sits at the very heart of leaf biology.

Carbon dioxide. Water. Both are necessary for leaves to spin sugar, yet getting one always risks the other. Leaves must inhale carbon dioxide through microscopic pores, stomata, on their surface. In contrast, leaves need to suck up water from the soil via complicated pathways of root hair and xylem tissue. Here's the problem: when leaves open their stomata to inhale carbon dioxide, they can't stop the water inside their tissues from evaporating outwards. Carbon dioxide in, water out.

Nothing comes for free. A phrase I might expect from my mother's mouth. It's also biology's byline: every biological form, every metabolic pathway, is constrained by trade-offs. Risk freights every molecule of sugar, no matter its sweetness.

His bulk, settling on the floor beside my bed, wakes me. It's Charlie, home from the bar, a pizza box in one hand, and the stench of beer on his breath. The bunk above me is empty – my sister, Laurie, is staying the night at a friend's house.

I hear: "Lyn, move over, let me lie down. I brought pizza."

Before I can even open my mouth, a verbal steamroller – one like I've never heard before – roars through the house, into my bedroom, shoving Charlie away.

It's my mother, outside, screaming at the top of her lungs, "Charlie fuckin' McInturf!" It sounds like she's out on the road. Did Charlie leave her at the bar? Somehow, I know she's running. Charlie knows too. Without saying another word, he stands and staggers from my room.

The front door bangs open. "Charlie fuckin' McInturf!"

In the dark cocoon of my bed, as the fight rages in the next room, and the greasy smell of pizza lingers, Charlie's appearance beside my bed is a boil beginning to fester.

Monday morning, sitting next to Laurie on the school bus, it bursts.

It's not long after, maybe a day or two, that I'm lying on my bunk reading, trying to ignore the argument picking up steam between Mom and Laurie. I want this fight to end. And then I hear it.

"Well, at least my boyfriend doesn't try to crawl into Lyn's bed."

The silence is suffocating.

My mother appears at my bedroom door and asks me if it's true. When I tell her yes, she tells me I better get up, there's packing to be done.

It's over. In the final exhalation of a sentence. No more late-night fights, no more marriage. Before the end of the day, before Charlie returns from his job on the Burlington Northern Railroad, his truck rumbling down our long drive, all of his clothes and tools are packed in cardboard boxes and garbage bags stacked outside the carport. From the door, my mother turns Charlie away.

I never see him again.

NOT LONG AFTER THE SWITCHBACKS, Marc, Maggie and I reach our lunch spot: a broad shoulder overlooking Okeover Inlet, with its smattering of conifer-clad islands lying far below us. We eat crackers and cheese and pepperoni in the shade of a leaning arbutus. Lunch over, Maggie and I scramble upward. At the top of the knoll, an ancient Douglas fir contorts more horizontally than vertically. I lean against its thick bark while Maggie goes off to pee.

No birds sing, no clouds drift.

The entire world, including me, feels pinned in stasis.

Yet, even as I listen for Maggie's return, I can't help but wonder how much of our impression of the world – weather, trees, mothers – is influenced by the limits of what we can perceive. The stillness of this moment feels absolute, but change the scale and *right now* transforms into interlocking circles of endless motion.

First, there is our Earth racing like a spinning top along a Hot Wheels track. Rotation within revolution, day within year. But all is not perpendicular. The axis of our Earth's rotation, relative to its plane of revolution, leans 23.5 degrees away from vertical. In each yearly orbit, first one pole, then the other tilts toward the light. Revolve cock-eyed around a star, and summer see-saws from one hemisphere to the other.

It doesn't end there. Rotate a globe with oceans and atmosphere in the warmth of a sun, and great eddies of air will rise and sink above its surface. Two fundamental laws of nature – the conservation of energy and angular motion – choreograph these swirling masses. Globally, the physics are complex: relative humidity dances with adiabatic cooling; the Coriolis effect swings convection and conduction; and radiant energy stomps heavy here, lighter there. Through it all, water is protean, changing from gas to liquid, even to solid, and then back again. But within each vertical eddy, the rules are simple: rising air packets, called low-pressure systems by meteorologists, spill water from condensing clouds. Sinking air packets, high-pressure systems, sponge up water, erasing rainclouds.[3]

Vegetation maps might label the Malaspina Peninsula as part of a temperate rainforest, but it's a rainforest that knows summers dominated by the descending air mass of the North Pacific High. Most years, the blue skies of this high-pressure system remain overhead from June to October. With little summer rain to recharge the soil, trees like the Douglas fir I'm leaning against will clamp shut their stomata by mid-July.[4] Today's heat may arise from our planet's cock-eyed path around the sun, but, collectively, revolution and rotation have stilled the breath of trees.

No wonder it's so quiet.

And then Maggie breaks the silence – crashing back toward me with the cracking of branches dried into brittleness. I think of my mother on her deck, surrounded by the calm of our absence. Daughters and mothers, mothers and daughters. Surely, their relationship is no less complex than those between climate and forests. Yet, as Maggie and I start downhill, it's clear I understand one type of relationship far better than the other.

From our lunch spot, the trail drops sharply. Three switchbacks later, we walk alongside a small stream. The ground squishes under my boots, and the air on my skin feels moist. I stop to look up. In the canopy, the deciduous leaves of red alder and bigleaf maple nearly blot out the blue sky. Down here by the stream, where the soil is always damp, these trees will be able to risk opening their stomata as soon as this heat wave breaks. Up above, on the dry knoll of our lunch spot, the Douglas fir and arbutus trees will not be able to draw breath until the thin soil beneath their canopy is recharged with rain. Two months, I think, before those trees can expect the Aleutian Low to swing south; two more months of drought until the gray clouds of this low-pressure system stack up into endless rain.

Too little, then too much. A pattern common to the words of my mother, and the precipitation of this forest.

We are in the kitchen of Jerry's Saloon and Steakhouse, where Mom bartends and I waitress the dinner rush to pay the bills now that Charlie's gone. Earlier I'd left, supposedly to go home, but had gone driving instead, with my best friend, Jenny. Wandering the skinny roads of Lincoln County, the dark interior of her mother's Town and Country station wagon filled with the sounds of Jethro Tull, the two of us chewing from a bag of psilocybin mushrooms, giggling our way over cattle guards and past rocky cliffs. And then we'd returned to Jerry's – the reason now long lost under the memory of our careful, high-stepped entrance through the front door.

The moment Mom sees me, her lips compress.

Come with me, she spits, before turning for the kitchen, where cooling fat congeals in the deep fryer, and the sinks gleam dank silver.

Away from the bar, she explodes. Only one sentence registers.

"You're never getting to university doing this shit."

And my mouth, liberated beyond care, shoots back, "Well it's not like you'd do anything to help."

Before the sentence is even finished, I can taste its nastiness, and when I
see my mother's fist coming toward me, I'm almost relieved. My nose crunches,
the giddiness of mushrooms draining into two streams of blood.

WATER POOLS, SLIDES, and then drips from my skin as I step from Wednesday
Lake. I rub down, thinking that if I'll never completely understand my mother's
influence on my life, then I'm nearly as mystified by mine on Maggie's. Arriving
at this lake, our destination for the day, Maggie had rushed for her bathing suit,
pleading with me to join her. I'd resisted, enjoying sitting still. But when I'd
changed my mind and shed my clothes, sliding naked into the water, Maggie had
squealed, "Mom, you're skinny-dipping! Can I?"

I'd been surprised she needed to ask, delighted by the relish with which she
slid out of her bathing suit. Porpoising through the dark, tannin-rich water, she'd
giggled, "It feels so different."

Who knew skinny-dipping with your mom could be cool?

Now I sit above Maggie, watching the reflection of the surrounding forest
shake and shatter in her splashing. For nearly 400 million years, forests have been
experimenting with how best to canopy the world in green. Behind the lake, the
hillslope is coloured in an evergreen canopy of cedar, Douglas fir, hemlock and
scattered arbutus. The leaves of these trees – scaled, needled or broad – will taste
four, five, maybe even more, summers before they die. It's not permanence but
unsynchronized leaf death that keeps these trees evergreen.

Down by the stream, the deciduous canopy of red alder and bigleaf maple
won't last much past October. Thin and flat, and packed with chloroplasts, the
leaves of deciduous trees compensate for their abbreviated lifespan by spinning
sugars quickly. The leaves of evergreen trees – often smaller, always built of denser
cells, and normally coated in thick wax – rarely photosynthesize as quickly as
deciduous leaves, but their photosynthetic season doesn't end with summer. On
warm autumn or spring days, when deciduous trees stand leafless, the evergreen
trees of this forest will be slowly spinning sugars – persistent tortoises catching up
to the deciduous hares stalled by autumn leaf loss.[5]

October. Burlington, Vermont. My friend Doug and I drive the final miles back to the University of Vermont. We've just finished guiding a week-long walking tour in the Green Mountains. My time in this deciduous, northern hardwood forest is coming to an end: my master's research was submitted last week; the man who first asked me to marry him is meeting me this afternoon. Together, we'll leave for a three-week rock-climbing trip before setting up house together in the coniferous-clad Rocky Mountains.

But when Doug and I arrive at the university, there's no car, no man, waiting in the designated parking lot. As the hours tick past, I imagine the worst: crumpled bodies on a distant road, a climber peeling away from a granite cliff. Finally, Doug tapes a note to the door of the building and takes me home to the house he shares with his girlfriend.

Inside, Doug asks for my address book and starts dialing. Much later, he hands the phone to me. The voice of the man is cold and distant. He's not coming. Not ever. Without thought, I dial another number. Mom answers.

"Oh, honey. That bastard. Come home right now."

And I go. Driving across the continent – from Vermont to Washington; driving through tears and the season's first snowstorm and the public radio news of the assassination of Israel's Yitzhak Rabin. Driving west to familiar weather.

ON DAY THREE, we all find the rhythm of the trail. Marc and I are up early enough to follow the undulating path of a pileated woodpecker through the arbutus branches above us. Marc pumps lake water into our water bottles, and Maggie and I finish packing. Heading south, the trail winds up over Hummingbird Bluffs, and then along the roller coaster of Gwendoline Hills. It's another hot day, but Maggie's delighted with our lunch stop at the hiking hut on Manzanita Bluff. She wants pictures of first her, then her with her dad, and then her and me poking out the hut's second-story window.

Atop the bluff, we chew the last of our pepperoni and crackers. Beneath our feet, the peninsula falls off, exposing a panorama of the Lund lowlands, Savary, Hernando, and Vancouver islands. Scanning the blanket of forest below us, I can find the deciduous billow of red alder and bigleaf maple only in the watery seams of creeks, or the open patches of recent clear-cuts. There's no denying this view. In this land of summer drought and winter rain, evergreen leaves most often win the race.

Not long after the 19-km trail marker, I begin to recognize landmarks. We've been here before, most often in the winter rain, during those years we came in December. Marc, Maggie, me, alongside my sister's family – all of us seeking refuge from the pained silences inside my mother's house. When the chaos of our bodies and overlapping voices swamped her tolerances.

Too little, then too much. A pattern with consequences. But not necessarily wrong, or bad. Just a pattern that shifts the ecological playing field in favour of evergreen leaves, allowing them to dominate coastal forests from northern California to southeast Alaska. The Europeans who came to log the "Jungles" would have found little in common with the deciduous woods that first gave rise to the word "forest." Rather than a landscape inhabited by beech and oak and maple, they found one populated by some of the world's largest and longest-living trees. A forest unrivalled for its ability to store carbon in root and trunk, branch and needle.

My mother's kitchen. Less than a week ago. Most of the family plays bocce ball beneath the apple orchard. Alone with Mom, the two of us in the same room, I stumble for conversation before asking her about her own childhood. Mom tells me how her maternal grandfather – the son of Irish immigrants, who worked most of his life as a labourer in downtown Vancouver – gave her a set of great books when she fell in love with reading as a young girl. Her favourite memory of this man, she says, was when he arrived at our house just as one of our hippie potlucks was beginning. Dressed in an overcoat (in my imagination, it was boiled wool, a relic left over from the trenches of the First World War), he'd said to her, "Well, if I'd known we were having a party, I'd have brought a bottle."

I know this man was widowed when his daughter was only 12, but when I try to ask Mom about her mother, my grandmother, and the man she married, my grandfather, she snaps, "Oh for Christ's sake, they were nothing more than rednecks."

Too little, then too much. I know so little of my mother's childhood. Where was my grandmother, I wonder, when my grandfather beat my mother with his fists? When he forbade her from attending university? When he banished her for smoking pot in the early '60s?

All these years, has my mother's inability to hold her tongue been a reasonable – one might even say an adaptive – response to life with a mother who said too little?

ARRIVING AT MOM AND BILL'S DRIVEWAY, Marc and I send Maggie ahead while we stow our backpacks and Freya in the basement. When we arrive in the kitchen, Maggie's already at the table with a frosty glass of root beer in front of her.

As I walk in, Mom turns to me, "So I hear you took Maggie skinny-dipping. Hell of a long walk just to get naked."

Without the rest of my family at the table, dinner is less boisterous than those earlier in the week. Bill serves homemade spinach spaghetti and prawns in a cream sauce. Knowing Maggie's fondness for both prawns and cream, I worry that her exuberant serving will raise the sharp edge of my mother's tongue. But Mom seems not to notice. Soon Maggie excuses herself to go read in the basement with Freya.

There, in the long light of a summer evening, our conversation stutters and then settles into an extended examination of bird politics and satellite TV. How, in this place of cedar and arbutus and bright water, I wonder, has my mother's life become so confined to her kitchen and deck? It's not that I object to the birds: in July, black-headed grosbeaks and goldfinches, spotted towhees and Steller's jays, visit my mother's deck. Turkey vultures drift in on thermals, and Mom and Bill tell stories of the ravens that visit each winter. At least, I think, my mother takes pleasure in the birds; the world perceived through satellite TV inspires only rants.

The next morning, before we leave, old anxieties resurface. I strip sheets and wipe surfaces, even as I know it's impossible to predict what might trigger the next verbal downpour.

October again. My mother's and daughter's birthdays are five days apart, and this year, I'm grateful that Maggie remembered to thank her grandmother for her birthday present, a blues CD by a New Orleans trombonist. I've never heard of Trombone Shorty, but Mom says he can blow a mean horn, and Maggie loved getting such a grown-up present.

On the phone, distance ringing between us, I confess that, with Maggie running full tilt toward her teenage years, I finally understand the lure of verbal threats.

"Of course," I say, "I don't actually tell Maggie that I'm going to beat her black and blue and rip her lungs out. I just tell her that the phrase is on the tip of my tongue."

My mother interjects, "It doesn't matter. Maggie would remind you that you and Marc decided not to spank, so she has little to worry about."

She goes on, "But you want to, don't you?" Like she did so often in my childhood, she has me nailed.

AS WE HANG UP OUR RESPECTIVE PHONES, laughter still echoing, I imagine my mother in her kitchen: the ravens returned from their summer jaunts, the turkey vultures and songbirds long gone for the winter, the satellite TV clicking on, Bill starting dinner. Outside my kitchen window, blue-black clouds pile against the hillslope, the first of the October rains to make it over the mountains.

Today's laughter doesn't dismiss how she really did lose it back when she was single, and my sister and brother and I were hungry. Hungry for food, hungry for her. And how, sometimes, there just wasn't enough to go around. Nor does it diminish my sadness that today I hear my mother's voice most often and best on the phone. What I have no way to tell my mother, even as her voice lingers in my ear, is that I know she loved us, even as she lost it. That she loves us still. That love propelled her to nag Bill for the name of a good trombone player when she learned what instrument Maggie was playing in her Grade 6 band.

But maybe if Mom and I are as different as the proverbial tortoise and hare, it's because we can't help it. I think of the summer stillness born of earthly motions too vast for me to perceive; the unique collision of climate and topography that builds a forest of evergreens here, a canopy of deciduous leaves there. Maybe the point is to see my mother less as a physics problem that needs solving and more as a daughter who grew true to the climate in which she was raised. To understand that all daughters, even as they become mothers, carry within them the legacy of those who went before – just as a Douglas fir trunk carries rings of wood narrowed by the shade of its elders. In forests around the world, the success of any one leaf type varies with climate, latitude *and* history – sometimes in unexpected and surprising ways. The same, I think, must be true of mothers.

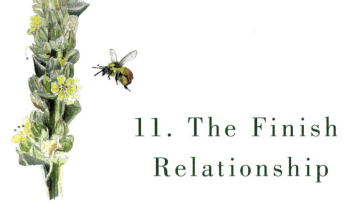

11. The Finish Relationship

IMAGINE THIS. Toronto, 1969. A young mother, not yet 25 years old but pregnant with an unborn son, hurries her two small daughters through the city. Bills, pilfered from a grocery allowance, are exchanged for a train ticket west to Vancouver, for a chance to escape a marriage gone violently awry. Not long after, the father will also return west, but by the time the children next see him, he will have become a stranger. In the years to come, the mother will rarely talk about the marriage she fled: the bruises, the beatings, the humiliation. But she will return often to the days she and her children spent traversing the continent by train – the young Black porter who came to sit with her daughters so she could visit the club car each evening. Eventually, this story will become the myth that calls their small family into being. That, unlike those families whose origin is rooted in ancestor and place, their beginning was the ride west across the grassy heart of the continent, when mobility served not to sever but to save.

Years later, when it's time for the younger daughter to travel east for college, her mother will convince her to take the train, to "see the continent." But after a childhood spent in the intermontane valleys of BC and Montana, the daughter will have little praise for the open solitude she finds on the flat heart of the continent. Instead, for three nights, she will try to imagine the man her mother fled years ago on that westbound train, but all she will see reflected in the window is her mother: tall and slim, long blonde hair piled impossibly into a beehive; one hand holding a glass of wine, while the other lifts a cigarette, its red tip a fierce glow against the dark.

ROADSIDE, HIGHWAY 97. Washington state. Late August, just past 5:00 p.m., with the day's heat still oppressive. I'm at the south end of the Okanagan River, a few kilometres north of its confluence with the Columbia. I lean against my truck, eating a stale egg sandwich I bought at a gas station, feeling my body jitter into place. Hot and tired, I scan the landscape, seeking something, anything, to shake off the desolation that has taken up residence within me. Battered and bony and dry, the view reflects my mood.

My day didn't start this way. Eleven hours ago, in the crepuscular light, I'd sat stock-still, one foot on the clutch, the other on the brake, eyes averted from the massive animals advancing toward me. Against the steering wheel, my hands clenched, inside my chest my heart jackhammered.

Two male bison: emerging from the morning mist, walking abreast down the centre of Yellowstone's Grand Loop Road, directly toward me. Two bulls, well into rut, identical to the dozens I'd watched earlier in the week slam into one another, over and over again, dust rising from embattled ground. Two bulls, each with a weight nearly the same as my little Toyota pickup, two bulls revved by the rut. Me, with no place to go on this thin line of asphalt, wedged between lodgepole pine forest and the Yellowstone River.

Five days earlier, I'd driven into the park for the annual gathering of my field journal group – a self-declared "litter" of boisterous women who have, for more than a decade, been my co-conspirators and mentors in field journalling. Together, we painted mud pots and geysers, tracked the stories of raven and goose through muddy sloughs, watched wolves and grizzlies wander through grasslands still green with rain. Through it all ran Yellowstone's bison.

Early yesterday morning, as we watched bison sweep first across the road, and then out along the broad curve of Hayden Valley, my friend Peg remarked, "They finish the landscape."

In the moment, I'd heard Peg speak artistically, but then, as the full depth of this valley filled my journal page, I wondered if what finished this landscape wasn't just bison per se, but the relationship *between* bison and grass. Little about bison, after all, makes sense without grass. Bison's tall teeth tolerate the erosive silica grains embedded in grass leaves; their complex, multi-chambered stomachs make space for the symbiotic bacteria who, unlike animals, can digest cellulose. Even their wool – curling boisterously around bull foreheads, hanging in goatees

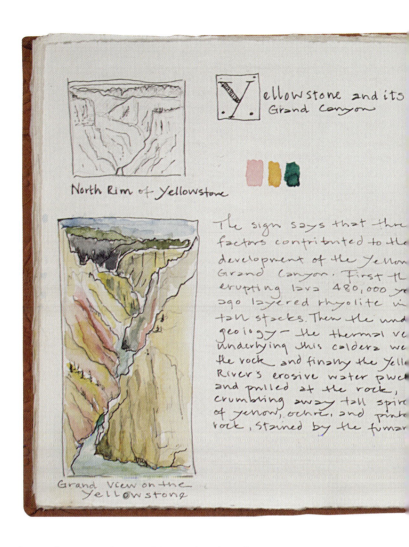

Yellowstone and its Grand Canyon

North Rim of Yellowstone

The sign says that thr[ee]
factors contributed to the
development of the Yellow[stone]
Grand Canyon. First th[e]
erupting lava 480,000 ye[ars]
ago layered rhyolite i[n]
tall stacks. Then the und[er]
geology — the thermal ve[nts]
underlying this caldera we[re]
the rock and finally the Yell[owstone]
River's erosive water pwe[lled]
and pulled at the rock,
crumbling away tall spir[es]
of yellow, ochre, and pink[ish]
rock, stained by the fumar[oles]

Grand View on the
Yellowstone

Vol. 20. Yellowstone and Its Grand Canyon/Hayden Valley

on males and females alike – serves as brooms to sweep 75 centimetres of snow from grass buried in winter drifts.

Likewise, without bison, the ecology of many North American grasslands loses coherence. Grasses co-evolved with grazers, and for the grasses that once sprawled across this continent's interior, no grazer has been more important than bison. Today, no one knows the exact size of the herd – estimates range upward of 30 million animals – that occupied North America before European settlement,

n Valley lookout.
ng for wolves, but I am
nced by the colours of early
g in the valley... we
the three grizzlies without
trying and then spend a long
watching / looking for wolves
for trees. Sketching bison in the
distance. Peg says, Bison finish the
scape"

but their loss remains a wound in the ecology of this continent.[1] Gone are the
hooves that once drummed and pounded trails of dust from Alberta to Mexico;
gone are the mouths that made space for less competitive plants, that grazed
the spread of grass rhizomes into being; gone are the bodies who distilled grass
into protein for grizzly and wolf, whose urine and feces and decomposing bodies
redistributed nutrients across the prairie.

Today, North America's bison herd numbers nearly half a million, but the vast majority live behind barbwire or within corrals. Only in a few places like Yellowstone has the relationship between grass and bison remained largely unfenced; have bulls been allowed to respond to the pull of the rut unmanaged. Yesterday, from a comfortable distance across the valley, the bulls had bemused me. Their oversized humps, their surprisingly dainty rumps. Their hormone-induced posturing atop Yellowstone's narrow roads that backed traffic into crowded bison jams.

All *this*, I'd thought, for fatherhood.

Bison and grass. Yesterday, I relished their relationship, but only this morning did I come anywhere close to understanding its physical reality. Two bulls: each with an enormous black head towering above mine, each with more than 1000 kilograms of muscle and bone to mock the resistance of my truck's metal and glass.

Sandwiched between the two bulls, some primitive instinct kept me looking away, avoiding eye contact, but I could hear them snorting, rumbling even, as first one, then the other, came even with my cab.

Finally, I'd snuck a peak in the driver's side mirror. One bull's rump filled the glass, while an abbreviated tail swished back and forth above a shockingly large brown testicle sack. One haunch stepped forward, and then the other. Truck bed... rear tire...tailgate...past. I'd exhaled the breath I'd been holding and put my truck back into gear. As the bison disappeared behind a curve, something like truth reverberated from the drum of my chest.

This, this quiet moment of terror, felt more real, more honest, than yesterday's bemusement. Bison might be integral to the grasslands of this park, yet engaging in relationship with the bulls' masculinity is never risk-free. Not for the grasses they graze, nor for the wolves and grizzlies they feed. And certainly not for the naive tourists who, each year, end up gored, catapulted or trampled when they crowd too close to the bulls, mistaking their momentary placidity for something it's not.

But the adrenaline rush of my encounter with the bulls evaporated hours ago. Since then, I've driven north by northwest: first up and out of the Yellowstone Caldera, down along the grassy bottoms of western Montana, up across the high, thin neck of Idaho's Panhandle, before sliding down the west side of the Rocky Mountains out onto a flatness – the Columbia Plateau – that seemed, in the heat of middle afternoon, to extend beyond Earth's curve.

I should've known better. I've travelled this plateau more times than I can count. Never once has it been easy, and today is no exception. This plateau is the northern reach of the Sagebrush Sea, an ecosystem whose shrubs and grasses once stretched intact across the basins that separate the Rockies from North America's coastal ranges. On this plateau, in late August, the temperature verges on the desolate. Worse yet, after a week in Yellowstone, my eyes have been recalibrated to expect native, not cropped, plant species. From Spokane west, the hours have unrolled in an unrelenting liturgy. Mile after mile, field after field. Winter wheat, for sure. Hay and maybe barley too. Millions and millions of lives dependent upon the machines that seed, fertilize, irrigate, thresh and harvest them. Millions of lives known only in aggregate. Millions of lives grown at the expense of those beings that once rooted the Sagebrush Sea.

And then the Grand Coulee Dam: more than 500 feet tall and nearly a mile long. Driving by, I was torn between looking and not looking. Between deciding which was more exhausting – the wall of cement ponding the mighty Columbia, or the ecological vulnerability that comes when we humans fray a landscape's connections to only those that include us.

But I also know that there are remnant islands of this Sagebrush Sea where its original relationships persist. In technical references, the names given to this landscape – Pacific Northwest bunchgrass, cold desert, sagebrush steppe, Palouse Prairie – are all words used to distinguish this ecosystem from the grasslands east of the Rockies. My home in Kamloops lies on the far side of an international boundary, but its ecology is far more closely tied to the rhythms of this plateau than it is to the montane grasslands I left in Yellowstone. As I finish my sandwich, I debate driving on with finding somewhere to camp; the possibility of sleeping in my own bed versus waking to a Columbia Plateau morning. At least, I think, I'll get north of the border before I stop.

THE OKANAGAN VALLEY: pronounced the same but spelled differently on either side of the international boundary. South of the border, I have little history with this valley, except as part of the drive between Montana and BC. North of the border, this valley holds nearly all the memories I have of my Canadian childhood. A childhood that was destined to end when, one September day in 1977, my mother gathered my sister and brother and me together to ask, "Do you want to meet your father?"

Not a question I knew how to answer at age 11. Not when our biological father had been the boogeyman Mom used when we pushed too hard. As in, "If you don't straighten up, I'm going to send you back to your father."

But, that fall, my mother, recently separated from her second husband, Simon, was struggling to support us on the wage she earned bartending part-time at the Armstrong Curling Club.

How much would be different today, I wonder, decades past her question, if we'd just said no? But a *father*. Until then, the only father-like figure I could remember was Simon, a rock 'n' roll guitarist who worked at the local sawmill to help pay the rent.

BACK ON THE HIGHWAY, the distant blue hills of the Okanagan Highland to the north hold the promise of fewer fields. With my Yellowstone-calibrated eyes, it's easy to wish for a bison or two. As I drive north, through the small towns of the American Okanogan – Malott, Okanogan, Omak, Tonasket – I think how tempting it is to simplify the complexity of landscapes or families into recognizable icons. The Okanagan Valley might look like cowboy country, but it's no Great Plains. Grasses of the Great Plains, as well as those in Yellowstone, receive enough rain to remain summer green. Grasses of the Sagebrush Sea, caught as they are in this region's extended summer drought, turn dormant brown by July.

Culturally, families are expected to have fathers. Biologically, fatherhood encompasses multiple shapes: even among our close primate relatives, some fathers contribute little more than sperm; others defend their offspring from predators but rarely feed or carry or teach their young; a few do it all, co-parenting alongside mothers. But, in any species, fatherhood, as a relationship, arises both from the social dynamics of a species and from the landscape in which it exists. Today, evolutionary ecologists believe that the high involvement of human dads – unique among the great apes – may have evolved, along with bipedalism and tool use, only with the rise of grasslands in eastern Africa.[a]

In a cooling and drying climate, forest trees diminished, and by 2.5 million years ago, our primate ancestors inhabited open grasslands. The calories of this landscape did not ripen in the fruits of trees but rode in the flesh of grazers, or hid protected in nuts and tubers. Biology predicts that any behaviour, including fatherhood, will evolve to that which best supports the survival of offspring. In Africa's new grasslands, children survived best if their fathers cooperated both with their mothers and other fathers. Hunting with other males acquired animal protein. Investing that animal protein back into their children's diet complemented the carbohydrates of tubers and nuts gathered by mothers. Human fatherhood, it turns out, has always been rooted in a web of relationships.

North of Oroville, the last town in the American Okanogan, I pass apple boxes piled along the road, orchards stepping down toward Lake Okanagan. At the border, there are only a few cars in line. When it's my turn, I drive forward, rolling down my window as the border guard steps outside the booth to take my passport.

As I respond to the familiar questions, the configuration – an open car window, one person standing, the other sitting – jolts me back to the first memory I have of the man who donated half of my genome.

IT WAS HIM, not me, sitting in a car. Me, standing, not sitting, holding the collars of our two over-sized dogs, both barking, tails wagging.

Minutes before, I'd been inside, waiting; my older sister, Laurie, in the shower; my little brother, Davey, just turned 8, playing in the backyard. I'd been the only one to see the slate-blue Mercedes Benz convertible turn into our driveway. A driveway that, up until that moment, had only ever hosted Volkswagen vans, sagging pickup trucks and clunky sedans.

When I went outside, the car window purred down and a man asked, "Do you know who I am?"

The man had red hair and a high, steep forehead – like mine. His face was freckled – like mine.

"My father," I replied before bolting.

I'M THROUGH THE BORDER north of Osoyoos, and I have a decision to make: go west now along BC 3, or follow Highway 97 into the North Okanagan before heading west.

At the intersection, I go west, into the dying light. The two routes differ by only a kilometre or so, but there's more intact grassland this way. Partway up the long hill, climbing out of the Okanagan Valley, I realize that, in 1977, this would've been the road my father drove – albeit in the other direction – when he came to meet us.

He hadn't been supposed to show up like that. My mother, still wary after nearly a decade without contact, had arranged through intermediaries to meet him several hours south of our home in Armstrong. But a minor car accident had delayed her, and, growing impatient, my father had decided to drive to nearby towns and ask after families that resembled ours. Or, at least, that's the story he eventually told us.

I knew none of this when instinct drove me inside, bellowing for my sister. I'd run into the house, out the back door, grabbed Davey and dragged him inside, locking the doors as I went. Leaving Davey in the kitchen, I'd flown into the bathroom and ripped back the shower curtain.

My poor sister: 13 years old, Laurie stood naked and wet beneath the pounding water as I declared, "He's here. He's not getting out of his car. I think he's afraid of the dogs."

RICHTER PASS. A low saddle connecting the Okanagan Valley with the Similkameen Valley. This evening, it gleams with tawny grass and blue sagebrush. Just beyond the pass, a narrow dirt road leads uphill to the South Okanagan Grasslands Protected Area. For years, this region hosted big, bold "No National Park" signs.

For more than a decade, the proposal for a national park in this valley has been described as pitting ecologists against ranchers. But, in BC, both ranchers and ecologists want more grass.[3] Grasslands make up less than 1 per cent of our land base, yet this tiny sliver of habitat hosts more than 95 per cent of our livestock grazing and one-third of our threatened vertebrates. There's no denying that the two possibilities – *range* versus *park* – imagine different relationship between people, grass and grazers, but there's also no denying that both carry the imprint of norms learned elsewhere. In the 1860s, European settlement of this valley was rooted in the assumptions of colonialism. I'm sure Frank Richter, the man who established the first ranch in this valley, with 42 head of cattle, wouldn't have heard the slur when, more than 100 years later, environmental historians called cattle "creatures of empire."[4] Likewise, the first national park in North America, Yellowstone, was built upon a romantic vision of the world – a vision found more in the writings of William Wordsworth, in the paintings of George Catlin, than in a deep understanding of the evolutionary relationship between grass and their grazers.

Rancher or ecologist: I trust neither (including myself) in the abstract. I know how easy it is for any of us to be pulled by what we *believe* to be true. No, in this landscape, my trust lies with those – primate or ungulate, seed herbivore or fungal mycobiont – who have apprenticed with its grass. Those who can distinguish the open hesitation of blue-bunch seed heads from the overlapping abundance of rough fescue's; who know the unexpected spiralling of *Stipa* seed that drives them into the soil, the lazy droop of Richardson's needle grass flowers. Those who understand the vase-like form of these bunchgrasses – unlike the rhizomatic spread of Great Plains grasses – as the architecture that risks growing stem tips well above the soil in order to funnel water to thirsty roots. Those who, in the early morning, are calmed by the slanted shadows stretching from one bunch to another.

I trust those who have worried through the complex equations of soil, grazing, elevation, aspect and fire that always mediate relations between bunchgrass and its most common neighbours – humped and gnarly, big sagebrush, complicated, if low-lying, biological crust. I trust those who scan hillslopes each spring for the first sign of greenup, for the reassurance that last year's grazing wasn't too much; those who know to count May's green trill of vesper sparrow, meadowlark and Say's phoebe against August's stark quiet. Most of all, I trust those who understand that the vulnerability of this grassland, just as with any family, can arise through either subtraction or addition.

OTHER THAN HIS ABRUPT ARRIVAL, I remember little from the day my biological father reappeared in my life. I began that day fatherless, yet secure in the extended hippie community living in the North Okanagan in the late 1970s. Little did I know, as we finished the day eating dinner at a Chinese restaurant, how meeting my father would cost me my home.

But how could I have guessed? This man's entrance into my life carried not just the emotional promise of a *dad* but the blunt power of wealth. Here was a man whose townhouse complex hosted an indoor pool, who drove a sleek Mercedes convertible, who owned, of all things, a chain of pet stores, who took me and my siblings on a clothes shopping spree when he flew us to Vancouver at Christmas. Years later, I still wonder about my lack of caution. Was it driven by a 1970s sense that *complete* families had fathers? Or did my longing for the material goods this man's bank account could supply drive my ready agreement? Either way, I should've known better. Mobility had allowed my mother to escape my father a decade earlier. If this time she feared not my father's fists but his money, her response would be no different.

It was the same Christmas we shopped with our father that my mother abruptly married an American stranger named Charlie McInturf. And less than a month later, we left BC, leaving no forwarding address, driving south by southwest across the Columbia Plateau to Montana, where we shared a three-bedroom rancher with Charlie, his ex–old lady, her four children, and three, sometimes four, other adults. Within six months, the inhabitants of this house would thin to just five – my family and Charlie – but my sister and I would not quickly forgive our mother either her new husband or our new geography.

DESCENDING INTO THE GRASSLANDS of the Nicola Valley, I'm one mountain pass away from Kamloops. If I was tired when I ate my egg sandwich in the American Okanogan, I'm beyond exhausted now. I need caffeine. Pulling into the Starbucks just off the highway in Merritt, I understand, once again, the comfort of standardized norms. See the green and white sign; know what to expect. Tonight, I'm relying on it.

Back in my truck, I drive uphill, out of the Nicola Valley, imagining the broad swoop of grass that extends beyond my headlights. Yesterday, standing in Yellowstone, I'd understood why conservationists have been working to unfence more of North America's bison herd. In both Canada and the US, in preserves small and big, bison now graze ground that had not felt their hooves in more than a century. Today, I couldn't help but carry this hope westward.

But what will restore the grasslands east of the Rocky Mountains will not do the same for the intermontane grasslands: both the form of their grasses and the pattern of precipitation argue against it.[5] If the growing tips of bunchgrasses are repeatedly grazed, their leaves are stunted and they have no horizontal rhizomes through which to regain lost growth. This region's extended summer drought means that in July and August there's little grass for bison, including freshly weaned calves, to eat. In this landscape, the relationship between grass and grazers has only ever been a seasonal dalliance, with elk and deer seeking greener grass at higher elevations as the summer progresses. Today, most ranches emulate a similar pattern, herding cattle through the lower, middle, upper grasslands, even onto forested range, before returning back downhill in the fall.

Years ago in Montana, my sister's relentless hostility toward my mother's third husband would win her permission to call our biological father. The sleek Mercedes Benz appeared in our Montana driveway, and my sister went away. When school ended that spring, I followed. But shared genes, alone, do not sculpt strangers into fathers and daughters. Late in the summer, after a prolonged negotiation, my father agreed that I could spend a final week with old family friends – the Morlands – in Armstrong. But then, two days after I left, he changed the locks on his townhouse and stopped answering the phone. My sister was refused entry. The last thing my father ever bought me was a plane ticket back to Montana.

Later, I would be told this man was addicted to Valium; that my sister's and my teenage bickering was too much for him. Later, my sister would tell me that when her new boyfriend and our father had fought, our father called the police. But, in the moment, as I learned from a stranger's voice on the phone that I wouldn't be returning to my father's house, I couldn't decide who'd betrayed me more: the man I'd been calling Dad, or my own simplistic expectations.

Yet, 40 years later, driving home to my own daughter in the dark, I feel nothing but relief at this man's actions. Who would we be today – me, my brother and my sister – if we'd spent our childhood alongside this man's moods, if we'd fallen in love with this man, who could bolt his door against one of us, who refused to answer his phone when we called? This is the truth of my family's origin: when my mother fled my biological father, she didn't just save herself, she saved all of us. And, maybe, the greatest gift my biological father ever gave me *was* that final plane ticket home.

Tonight, as the road climbs to just under 1500 metres – the Surrey Lake Summit, the last before home – I try to imagine the sights and sounds that accompanied the arrival of the first cattle in this valley. The creak of stirrup, the thud of hooves. I wonder if the men riding alongside spoke with an American twang, or with an Englishman's droll; were they headed northwest to the Caribou gold-fields, or northeast to the ranches near Kamloops that are still famous for their long-ago polo matches and tea soirees? Most of all, I wonder if they had any idea of the trouble that would follow. The 50 years of overgrazing that would fragment the lower grassland's biological crust, shredding its microscopic bodies of algae, lichens and moss, and filling this range with a "dark black dust."[6] How recent arrivals, like cheatgrass and knapweed, would find places to root in an ecosystem denuded of its native grasses, stripped of its nitrogen-fixing, soil-cementing bio-logical crust. Did they understand how their arrival, in the words of writer and ecologist Don Gayton, a man who has done more apprenticing to grass than most, would result in nothing less than "a full-scale ecological conversion"?[7]

In my family, I don't have a father; I have a *Bill*. By the time Bill – now married to my mother for nearly 40 years – entered my teenagerhood, the space normally occupied by a father had been overgrown. Born in the open sagebrush of eastern Montana, Bill – woodcarver, forest surveyor, economic developer – courted my mother in the rocky valleys of western Montana. Together they moved west to the coastal forests of Washington, and then, finally, north to BC. Today, my relationship with Bill falls into no easy taxonomy. Stepfather is too removed; father belies our particular history. I don't doubt my relationship with Bill; its strength is measured in the frequency with which *Mom&Bill* – one word, one entity – crops up in conversation. My hesitation to call this man "father" or "stepfather" stems, I think, from my reluctance to crowd the rich and lengthy relationship I have with my mother's fourth husband into something it's not.

I am sliding down toward the lights of Kamloops: home is just minutes away. Tomorrow, I decide, I will take my journal back to the grasslands I know best: the swoop of grassy hills surrounding Botany Pond, just northwest of Kamloops where the rough fescue mounds, and the mariposa lily capsules split three-parted.

If the cows are already there – grazing their way to their winter pasture in the valley below – I won't call this grassland something it's not. Neither cows nor I finish this landscape: our relationship with its species is still too shallow; our collective threats too great. Yet, in both landscapes and families, failed relationships help point to those that might succeed. A century ago, overgrazing reduced this grassland to little more than cheatgrass and dust. Now, each May and June, its blanket of grass throbs with green, wildflowers bloom between wheatgrass and sage, bear wander its folds. Walking in tomorrow, I will know that, even as the cattle risk crust, they preserve open space. I will remember the legion of ranchers and ecologists who spent decades negotiating the careful pasturing and long rotations that slowly allowed this grassland to reknit. Sitting down to draw in my field journal, I will worry. I always do; worry, after all, is part of relationship's finish.

CONSIDER THIS: Highway 5. Just outside Kamloops. A woman drives down toward the lights of a small city that sits at the confluence of two big rivers. Just past ten in the evening – her daughter will be asleep, but her daughter's father, her husband, will be waiting up for her. She knows that once she arrives home, she will keep this man up even later, needing to tell him of her morning encounter with two bison bulls. Needing to tell him how she was pinned between the massive animals, how one testicle sack nearly filled her driver's side mirror, how the day unfolded in the long, slow drive – one their own small family has made multiple times – from Montana to BC.

If the woman in the white truck was disappointed with fathers until her husband became one, she's glad she never gave up on their possibility. For the last nine years, since the moment he saw the crown of her head emerge, her husband has shared her daughter's life. It has been a fatherhood built not, like that of bison, upon big head-slamming dramas but on little moments: father and daughter walking hand in hand down the street; the father's intercession between the woman's impatient words and the daughter's imperiousness; the lazy afternoons in his garage-turned-woodshop – the daughter reading a book, him slotting wood. Today, neither the woman nor the daughter does well in the father's absence.

But if both landscapes and families are finished in the vulnerability of relationship, the woman also knows it is a history of shared moments – big or small, positive or negative, negotiated or not – that transforms strangers into kin. Sliding toward home, the woman knows that, tomorrow, she will walk the grassland of Botany Pond. If she's lucky, her husband and young daughter will walk with her. She will know that she comes not just for the plants but for the metaphors, that *this* is the landscape that has helped her make sense of the startling exceptions in her life: all grass is grazed except when it's not; all families have fathers except when they don't.

Is it any wonder she wants the touch and texture, the species and community, of this grassland to be woven deep in her daughter's memory?

12. The Collecting Basket

Toronto, 1968: In the black and white photo, my sister Laurie holds my hand, facing the camera with an enigmatic smile. Together, our bodies intersect the outlines of a tennis court. Neither my sister nor I wear shoes, just socks. Hers reach to mid-calf; mine pool about my ankles. On my sister's body, a flowered sweater hangs mis-buttoned above smooth shorts. Thick hair surrounds her round face – 4 years old and still without scars. Only 2, I occupy my usual position beside her. A pleated dress clenches my chest, and my wispy hair is already pushed back into its signature cowlick. Collected with the click of a shutter, chemistry, not memory, has preserved this moment from the past.

COLLECTING – the parsing of the desired, known or not, from the undesired – lies at the heart of my chosen discipline. The first botanist I worked for gathered his specimens into a tall Adirondack pack basket woven from strips of black ash. All through a humid Vermont summer, I followed his basket beneath a deciduous canopy, imprinting on the warp and weave, on the traditions of a science.

Today, there's no denying I'm a collector, but I've long wondered about my allegiances. Why plants and not beetles? Or butterflies? Part of the answer, I suspect, lies rooted in the identity of those I gather most. Plants whose biology turns where they live into an oversized collecting basket. One that collects without human hand, that preserves without intent, that gathers specimens for millennia without fatigue. I know I visit such a place each fall.

Called Placid Lake, this basket is found on a broad volcanic plateau just inside the boundary of Wells Gray Provincial Park, just two hours north of Kamloops. From the air, it appears as an irregular, dark ellipse surrounded by a ring of shrubby vegetation, and on the topographic map it is annotated with the symbol mapmakers use to indicate the presence of wetlands. Technical guides label Placid Lake as a poor fen, and ecologists classify fens, as well as bogs, as peatlands.[1]

Peatland – roll your tongue over the name, taste its strangeness. Land, you recognize. Land, you can step onto. But peat trembles underfoot, less well defined. It's what the dictionary calls "vegetable matter partly decomposed." Mostly plants, but some animals too, caught in a wrack of delayed decomposition. Strange in the mouth, even stranger in the field.

Walking on the shady trail, nothing prepares you for the abrupt openness of this place – especially not on a late September day, when white clouds reflect in dark water and the leaves of bog birch gleam with gold. The keystone species of Placid Lake is neither tree nor flower – the typical masters of plant biomass and diversity in BC – but is, instead, a moss called *Sphagnum*. It is the life and death of this moss that has built a landscape where the ground undulates in hummock and hollow; where round leaves of sundews announce their carnivorous appetites with brilliant, but sticky, red glands; where the roots of black crowberry and bog laurel are stitched into place with the threads of symbiotic fungi; where the astringent odour of sun-warmed Labrador tea fills the air, and red-brown dragonflies buzz lazily. Shaped by the wrack of *Sphagnum*, this landscape subverts even my most basic notions about death and decay.

Each September, just as the bog cranberries ripen, I bring second-year botany students to Placid Lake. As we step from forest soil to peat, the unique botany of this place never fails to astound the students. But this peatland is more than just its plants. It's a physical accounting of change; a record book that first opened more than 10,000 years ago. When Pleistocene glaciers were abandoning this valley, a chunk of orphaned ice – approximately the same size as this peatland – was smothered by the sand and gravel sluicing off the melting glaciers. Long after the glaciers had retreated, the orphaned ice block melted, collapsing into a pool of water surrounded by gravel.

gaultheria hispidula
·closely related to eastern tea-berry.
methyl salicylate is active chemical, closely
relatd to aspirin's active ingredient. edible
berries

oxycoccus oxycoccus
· all native peoples gathered berries. the name
cranberry may be a corruption of "crane-
berry" because head+stalk resemble crane.

ledum groenlandicum
· across entire continent, leaves were used
as a stimulant tea. contains alkaloids
called andromedotoxins, toxic to livestock.
leaves boiled to make aromatic tea. must
be boiled for a long time to destroy alkaloids

kalmia microphylla ssp. microphylla
opposite leaves, contains extremely poisonous
alkaloids. 10 little bumps hold to anthers
that are spring loaded. to throw pollen on
visiting insect.

empetrum nigrum
· can make beer or sparkling wine, favorite food
of bears. name derives from "en petros" Gk
for "on rock."

rubus pubescens
· rubus is a poorly defined, rapidly hybridizing
complex — poorly defined.

Placid Lake
notes

empetrum
nigrum

little brown
mushroom

sphagnum

the plan
t

Wells Gray Vol. 1: Placid Lake Notes

Make a depression, and it will collect. Chance and gravity make it so. But preservation requires more. Placid Lake began to subvert the chemistry of decomposition when its shoreline was colonized by *Sphagnum*. Just as beavers can still streams into ponds, *Sphagnum* can invert lakes into elevated peatlands. Water wicks upward along its stems, and into this rising water *Sphagnum* releases hydrogen ions, acidifying the lake beyond the tolerance of many decomposers like bacteria and fungi.[2]

In peatlands, decay does not follow death, at least not completely. Instead, the alchemy of *Sphagnum* slows decomposition, allowing more peat to accumulate, water to rise, and pH levels to lower. Over millennia, peat builds outward from the shore, cinching closed pools of open water. Buried within the peat are relics – big and small – fossilized by the embrace of anoxic, acidic water. In 1950, peat

grazed betula glandulosa.

drosera rotundifolia
• insectivorous plants, along the leaf edges are
gland tipped hairs that produce drops of red-
coloured, sticky juice. insects, attracted by
juice adhere to it and are trapped when the
tentacles and leaf edges curl in around them. leaves
then secrete digestive juice. juice has reputation
for curdling milk. sap also contains an antibiotic
that is effective against bacteria; used to treat
tuberculosis, asthma, bronchitis, and coughs.

ledum groenlandicum

gaultheria
hispidula

september 14, 2007
with the 2007 211 Class

inity of Sphagnum

sketched from photo of
floating dock

aerial view of placid lake,
blue water surrounded by a
vegetation

harvesters pulled a 2,000-year-old corpse from a Danish peatland, the man's skin remarkably intact beneath the noose around his neck.[3] No human corpse has ever been found within Placid Lake, but its collection includes microscopic plant pollen, small bits of sedge and alder, carcasses of insects, maybe even the flesh of a mammal or two. Bodies linger in the northern lands where *Sphagnum* grows.

I tell these stories each autumn as my students and I walk into the lake. What I do not tell them is how I linger over the idea of this peatland as a collecting basket. David James Duncan used the term *river teeth* to describe the "knots of experience that once tapped into our heartwood, and now defy the passing of time."[4] Perhaps it's the bias of my discipline, but I keep sensing these "knots of experience" as a layered wrack, caught by chance in the peatland of my own history. There's a funny thing about peatlands. Near the edge, the peat extends all the way to

The Collecting Basket
—

the bottom, but close to the centre, the peat thins into a floating mat. Out there, it doesn't take much to ripple the surface. And if you bounce too hard, you can punch right through. I know – I've done it, my left leg sinking up to the thigh before I caught myself.

Vancouver, 1969: A weekend morning. A single mother, not in her bed but asleep on the couch, university textbooks piled on the floor by her side. My sister and I, 5 and 3 years old, wander the upstairs of our creaky, two-story house. The memories are fragmentary. Some images may be real, or lifted from stories long mistold. A narrow hallway leads from our bedroom to the bathroom. I feel the cold slickness of worn linoleum under my feet. True or imagined, the white porcelain bathtub, with its clawed feet standing atop a patterned floor, remains fossilized in memory.

A pile of toilet paper teeters on my sister's lap. An illicit pack of matches. Years later, the sight of my little brother holding matches in his chubby hands would send my mother into an incandescent rage. But, in 1969, my mother lay asleep downstairs as the flames tore through first the soft toilet paper, and then my sister's nightgown. My bare feet slapping down the long hallway – my mother's bed empty. I hear my younger voice saying I didn't know water could equench fire. I that I only know to blow out matches, to blow on the fire licking up my sister's body.

The intervening years have mineralized these bits of memory. Others are decomposed, indecipherable. I have to imagine my sister's screams. Did her raw voice pull my mother upstairs? I have to imagine an ambulance, my sister carried downstairs on a stretcher. Did I stay with neighbours during the long silence after my sister was gone? That fire, that licking flame 40 years ago, was born into my family just as surely as my little brother was. As a family member, it has no name, but my sister carries it, embedded in the scars that crawl from her groin up to her chin, just as an older sibling carries her younger brother. Forty years ago, in the single strike of a match, my sister and I forever altered the trajectory of our family history. Laurie, chubby-cheeked and thick-haired, was relegated to the land of the permanently disfigured. I was not.

MANY YEARS LATER, this is what I know: the line between controlled and rogue fire is paper-thin. As a botanist, I know that all plants feed upon the fire of our sun, gentling its energy to spin sugars from water and carbon dioxide. Yet strike a match in a dry forest, let loose these stored sugars in a wilding fire of uncontrolled energy, and trees will be charred into black skeletons, flesh transfigured into ash. That same wilding release, controlled in our stomachs or the cells of a decomposer, generates the chemical reserves needed for life. Capture. Release. Capture. Release.

This is a cycle that in many places leaves little trace. Yet in Placid Lake the biology of *Sphagnum* short-circuits the release of energy, decomposition slows, and artifacts collect. If memory is a layered wrack, so too is a peatland. Within this wrack, one single artifact, isolated from others, says little. But a collection of artifacts structured in time writes history. We could find this history, if we were patient enough to core down into Placid Lake. Layer by layer, metre by metre, we could extract tubular samples of peat, each one archiving a package of time, each one containing relics of the characters that once lived and died in this place. After months spent bent over microscopes, our backs would ache, but we could chart our tally of fossils into a series of waves depicting the abundance of individual plant species rising and falling throughout the lake's history.

Such a chart would echo with the tales told by peatlands across the northern hemisphere. The oldest, lowermost relics are nearly always from the community of plants that can invade the raw landscapes left behind by glacial retreat; plants found today on the margins of ice, high up in the mountains or far to the north. In BC, the pollen of pine dominates the next layer, and would have been collected as these trees spread north from their glacial refugia.[5] Other species would lag behind. In most BC pollen cores, red cedar – that iconic tree of our wet forests – arrived only 5,000 years ago.

Author Robert Macfarlane writes, "landscape has long offered us keen ways of figuring ourselves out to ourselves, strong means of shaping memories and giving form to thought."[6] An ecosystem that defies ignition, built from organisms that can tame star fire into sugar and thwart the controlled burn of decay. Is it any wonder that this collecting trip into Placid Lake now forms the heartbeat, the metronome, of my yearly calendar?

Vancouver, 1982, 2004: Loud music blares at my sister Laurie's wedding reception, and I swing her first daughter in my arms. At 16, I want little to do with the ritual my sister is embracing, but I don't mind dancing with my niece. The Kitsilano Neighbourhood House is packed with people – a collection of characters from my family's immediate and more distant history. My sister's parenthood and marriage will predate my own by more than 20 years. Years later, when we are both mothers, my sister will drive through the Vancouver rain to sit in an austere hall as I defend my doctoral dissertation. My husband will leave the room to stroller our restless daughter over the university walkways. I will not remember the arcane questions asked and answered, but I will long remember my sister's presence. She will be seated just behind the panel of senior professors who are interrogating me. Her body will still carry traces of that long-ago fire, but I will no longer see them – the raised and ropy edges of her scars made invisible by familiarity. Memory will preserve only her attendance, standing beside me, figuratively if not literally, occupying her usual position in my life.

IF SOME LANDSCAPES help decipher history, some also imagine new possibilities. Stand on the peat, I tell my students. Revel in the strangeness that surrounds you. In Placid Lake, dead moss bears the weight of living trees. Plants sip nitrogen from insect bodies, not soil. Know the two sides of life. Capture, release. Scarred or not, all of us are caught in this cycle. But know also that cycles subverted can, in turn, transform.

In this place, new rises from old. Imagine, just for one moment, I tell my students, life as subversive *Sphagnum*. Feel the water molecules wicking between your cells, the ions pumped across cellular membranes. Sense the history massing underneath you, even as the composite of your body builds into hummock and hollow. Know the horror of a partially digested fly still hooked in a sundew's sticky embrace. Comprehend the push and pull of the strands that weave this place whole.

13. Wearing Red

IN LATE SEPTEMBER, as I drive away from Botany Pond, I'm already plotting pigments. Beside me, sealed into a plastic bag, is a collection of colours from this season: yellow aspen leaves, bright red kinnikinnick berries, a merlot-red Douglas maple stem, green Douglas fir needles, and a lone yellowish *Suillus* mushroom.

From experience, I know that no one watercolour will suffice, not even for a single berry or leaf. Instead, I will need to layer washes: first, Hansa yellow light, and then olive green for the Douglas fir needles. Maybe yellow cadmium and quinacridone pink for the kinnikinnick berries. Art mimicking biology: using the reflected light of watercolours to imitate the shift and swirl of plant pigment. There are rules for how watercolour pigments should be combined, but tonight I'm going to ignore them all. I want to mess about in autumn colours, regardless of outcome.

Rooted, I've come to realize, doesn't mean stagnant. Author Victoria Finlay writes that colour shouldn't be thought of as a thing, but as a *doing*, as waves of light bounce from one object to another.[1] Each plant in the plastic bag beside me is absorbing some wavelengths of light and reflecting others. Colour manifests when my eye intercepts the reflected light and my brain assigns a name – red, yellow, green – to specific wavelengths.

Today, outside my truck window, the colours are falling. As I drive, bits of autumnal gold and amber and yellow spiral through space, carrying colour from canopy to ground. This – a world dripping in gold, alive with falling pigment – will disappear in a few days, a week at most. Leaves on the ground will muddy; the branches above will empty. Here and now, I anticipate colour's transience. I wish I had known the same about young girls and colour.

NINE YEARS AGO, my daughter Maggie started kindergarten dressed in as much of the rainbow as she could fit on her small frame. Left to her own devices, she would dress herself in soft pink pants, two contrasting patterned skirts, and then match her bottom with the brightest tank top she could find, layered over an equally bright – but differently coloured – long-sleeved shirt. She'd walk out the door, hand in hand with her dad, her small body a skipping swirl of contrasting fabric and colour.

If not quite alarmed, Marc and I were a little startled to have produced an offspring with such a flamboyant sense of style – let alone the piles of laundry she generated. Like many field botanists, we're generally more comfortable in earth tones scavenged from a thrift store than anything you might see on the cover of *Vogue*. Eventually, we negotiated a family compromise: Maggie could wear whatever colours in whatever combination she wanted, but no more than four different garments at any one time, including underwear. Clothes, we handled. But then one Sunday afternoon, not long after she started school, Maggie bounced into our kitchen clutching lipstick and nail polish – both bright red – cajoled from her paternal grandmother. She should, she proclaimed, get to wear both to school. They weren't, after all, more clothes.

"Lipstick and nail polish are worn by women looking for husbands. You don't want a husband, do you?" was my *mom as biologist* reply, knowing Maggie considered boys other than her dad barely above pond scum.

Even now, years later, I remember Marc's stunned silence. Maggie, in comparison, was little fazed.

"But Grandma Sandy wears lipstick, and she says she didn't want another husband after Grandpa Bill Jones died. And my friend, Lisa – her mom wears nail polish and lipstick every day, and she already has a husband."

A pause, and then from Maggie: "Does Lisa know why her mom wears lipstick?" Abandoned by Marc's laughter, I was left sputtering – any coherence obliterated by the image of Maggie whispering to her best friend the *real* reason her mother coloured her lips and nails.

Not without triumph, Maggie left the kitchen, a silver lipstick tube in one hand and bright red nail polish in the other.

But then, over the course of Grade 3, Maggie shed her flamboyant wardrobe, colour sloughing off in discrete bits. In September, the multiple layers went. When the first snow fell, skirts disappeared. Just before Christmas, for the very first time in her life, she asked for jeans.

In two weeks, Maggie will turn 14 years old. Today, her wardrobe is bereft of all exuberance. This fall, she would no more wear red – lipstick or otherwise – than she would fly to the moon. Maybe that's part of the reason why I was so determined to find something wearing red today. Not red lipstick, but red pigment.

IT IS A FEW HOURS EARLIER, on the trail connecting Botany Pond with the trail system surrounding our local outdoor education centre at McQueen Lake. I know I should be hurrying, but my feet keep stopping, my body transfixed by the falling colour. Red leaves and autumn: in many places, they go hand in hand. But not here. The autumnal red-deficit in BC's interior arises partly from the dominance of conifers that stay green all year, and partly from the palette of our deciduous trees. Quinacridone gold, Hansa yellow light, even cadmium yellow or raw umber – *yes*. Red – *no*.

For years, I've short-changed autumn in Kamloops: certainly, my field journals record fewer entries in this season than any other. It'd be easy to blame this neglect on an academic schedule that is so out of step with the biological year. We gear up in September, just as the growing season winds down. But now I wonder if my seasonal snubbing stems from the years I spent in Vermont. How, in October, can the subdued hues of my western coniferous forest compare to the fierce fire that races across this continent's eastern deciduous forests?

RED LEAVES IN AUTUMN. It's one of the few botanical phenomena that can pique the interest of even the most recalcitrant botany student. I should know – I use it shamelessly in my second-year botany course each fall. Here's what never fails to grab students' attention: we know *how* leaves turn red in the fall, but we don't know *why*.[2]

Green and yellow, in comparison, are easier. These colours largely arise from pigments contained within chloroplasts, the tiny membrane-bound structures within cells that absorb reddish-blue light from the sun to spin sugars from carbon dioxide and water. In summer, leaves are weighted with so much chlorophyll that all we perceive is the green light that reflects from this molecule. But even when we don't perceive them, there are other pigments – carotenoids – reflecting orange and yellow alongside chlorophyll's green. We perceive the full glory of these pigments only when, in the final days before leaf fall, trees reabsorb leaf chlorophyll faster than they do leaf carotenoid. The result: an elegy of gold.

In comparison, the pigments that colour leaves red – anthocyanins – are never found in chloroplasts but are stored in sac-like cell compartments called vacuoles. In the light economy of plants, anthocyanins are expensive. Not only do they take energy to build, but their presence reflects the very wavelengths of light plants use for sugar making. Playing no role in photosynthesis, these anthocyanins are largely missing from summertime leaves. But then, in a surprising extravagance, some deciduous trees and shrubs anoint their leaves with these red pigments in the days and weeks before leaf fall.

Last week on campus, I gathered red maple and oak leaves (all non-native to Kamloops) to show students. Today, I'm trying to find those same pigments in native plant species – even if it means searching lower in the canopy. Most of the students I teach grew up less than a day's drive from Botany Pond, and I want them to be able to take home at least one of the stories they've learned in my class when they leave next week for Thanksgiving. Wouldn't it be good, I think, if this year's botany students, out for a post-prandial walk, bellies full of turkey, could regale their families with the mystery of autumnal red?

But the best red I've found so far fills only half a red-osier dogwood leaf. In the vocabulary of my watercolour palette, the red on this leaf is more carmine than cadmium red. More blue than orange. More warm than cold. Botanists – who have at least 30 different words for red – would call this leaf *purpureus*, meaning reddish-purple. I like "carmine" better. The word carries transience in its history; used in many 19th-century watercolour paintings, this pigment has since faded to brown.

In the field, carmine's a good red, but I want more than just half a leaf. I walk on, the rhythm of walking transforming my desire into a more general attentiveness. *Amelanchier alnifolia* – yellow carotenes; *Betula occidentalis* – raw umber; *Betula papyrifera* – yellow cadmium. Snowberries all white, beneath nutrient-starved, brown leaves. Above me, a flicker calls, and a red squirrel scolds. I crane upward until I find the squirrel's quivering indignation atop a Douglas fir branch. Behind me, a raven *carooks*.

Just me and the birds and the squirrels. I'm relishing the moss-covered shade of this trail, when abruptly I emerge into a sunlit opening. Shrubs crowd around a lone ponderosa pine snag, a legacy of the mountain pine beetle that decimated much of the pine in this valley several years ago.

Beside me, the stems of Rocky Mountain maple gleam merlot red, but the maple's remaining leaves are a dull yellow, spotted with brown patches of decay. Leaf red – I never thought it would be so hard to find. We had red needles when the mountain pine beetles came through with their ravening mouths. Red is dead, we said.

Out here in the bright sunlight, it's warm. I slip off my pack, dropping it to the ground, and strip off my sweater. Tying it around my waist, I reach down for my pack, and there it is: a carmine mixed with a dash of orange. The red leaf of wild strawberry, *Fragaria virginiana*. And there, further along the trail, bunchberry leaves all quinacridone pink and carmine. I dig out my camera. These are reds worth talking about. I only wish this year's botany students – all 60 of them – could be here with me now. There'd be no better place to talk about leaf red than this sunny opening created by animal mouths. Sun and predators. Defence against one or the other, it turns out, is our best guess for *why* leaves turn red.

Red against the sun. As a redhead, I've spent much of my life worrying about the sun. Harnessed, the sun's energy fuels photosynthesis; unharnessed, this same energy can damage epidermal cells, whether on a leaf or a face. My tolerance of the sun's energy changes little from one day to the next; leaves' tolerance, in comparison, varies with temperature, canopy openness and chlorophyll levels. Leaves, plant physiologists tell us, are the most vulnerable to sunburn early in the spring and late in the fall. The shade of an anthocyanin umbrella, this potential explanation goes, may improve trees' ability to reabsorb chlorophyll from dying leaves.

As a hypothesis, *red as umbrella* vies with *red as warning*. The logic for the second hypothesis goes like this: Rooted in the ground, plants use both physical (think spines and thorns) and chemical defences to protect themselves from being eaten. For those of us drinking a glass of red wine, anthocyanins can be healthy; for insects chewing on leaves, anthocyanins can be lethal. Plants in prime condition can boast of their chemical arsenal by literally painting themselves red – the plant equivalent of skull and crossbones. Over time, plants that bear the cost of producing anthocyanins suffer less damage as leaf herbivores evolve to avoid their red leaves.

Umbrella or warning: Which theory is right? To date, no one knows. There's evidence for both. But what I think about, as I pack up my journal and camera, is

Vol. 36: Lac du Bois

st that transitions
a mixed forest
slope. Mosses
ning and
taller here
and boletes

It's a tease, this wild colour that arrives and then is lost. Un-synchronized. A little here and then there. Grey wallows out from underneath knapweed even as Acnatherums turn to straw. Students scatter across the slopes. Some with a view. Others alongside the cabin foundation. Somebody's squawking in the pond even as the pale september sun squints my eyes. Raven crawworks, just once.

The pond is the most busy. Numerous mutterings that I can't identify. From here the ducks are hemi circles of interrupted light on a flat surface.

Higher along the McQueen Lake Day Centre Trail
   Cornus stolonifera   – merlot red
   Amelanchier alnifolia - red
   Betula occidentalis - yellow
   Betula papyrifera   – yellow
Snowberries all white beneath nutrient-sucked lvs
   Rosa sp w/ bright red hip

Flicker calls and chickadees too. A red squirrel makes urgent calls – from a branch crook. Suillus everywhere ; kinnikinik berries; lonicera lvs – in the pine-beetled opened stands

Rocky Mtn Juniper, Common

that the production of red leaves, regardless of the cause, reflects plants' fidelity to place. Unable to pick up their roots and shuffle away from mobile predators, or from a glaring sun, plants have become masters of biochemistry – painting not just red but an entire palette of colours across their bodies. In comparison, animals like us, with all our mobility, have remained woefully inadequate at manufacturing pigment. Only one of the biochemical pathways used to synthesize plant pigments is found in my body – albeit to make red hemoglobin not green chlorophyll.[3]

But just because I can't produce many pigments doesn't mean I don't get to play with them. Turning the far corner on today's loop, I stoop again and again to pick the colours of this forest, in this season. Arriving back at my truck, I dig out a plastic bag and seal the colours in. Not only will I paint the colours of this forest in my field journal, but this weekend I'll bring Marc and Maggie out to see them.

For once, I keep my resolutions. That night, I fill my field journal with paintings, and the next Saturday, I chide Marc and Maggie out of bed early. By 9:00 a.m., the three of us, along with Freya, are back on the trail.

Within 100 metres, everyone is miserable.

I talk too much, wanting Maggie to imprint on her home's version of autumnal red. Maggie, who has become increasingly annoyed by my tendency to lecture, could care less. Between Maggie's sullenness and my exasperation, Freya lunges and whines – all obedience training lost beneath the onslaught of fresh deer and bear scent – making it clear if we let her off the leash, we would wait a long time for her return. My normally patient husband – caught in a triangle of misbehaving females – finally loses his patience and hands Freya's leash to Maggie. "Run!" he snaps.

And they do. Teenager and dog delighting in the sheer physicality that is their animal inheritance. As a botanist, I rely on the mystery of red leaves to intrigue botany students: the multiple hypotheses proposed is the stuff of strong inference. As a mother of a daughter who turned away from colour, I spent hours in academic databases, trying to track down explanations for her behaviour with little success. Yet, like me, many of my women friends had similar stories about themselves, their daughters, or both. Maybe, I thought, my own preference for earth tones had less to do with a field botanist's aesthetic and more to do with a desire for invisibility. Maybe, I thought, we avoid colour to hide behind conformity of earth tones (but why can't we all be bright conformists?), or to reject the trappings of impending puberty.

But today, following in Maggie and Freya's wake, I remember that the link between cause and effect, colour and function, isn't always direct. All four of us on the trail today carry red inside our veins, but the colour of our blood doesn't improve its function. Blood's colour is a by-product of the chemical structure of the hemoglobin molecule that transports oxygen. If a slightly different chemical structure could latch onto oxygen, we might carry blood coloured violet or quinacridone gold in our veins. Then, when we stayed too long in the sun, our skin would burn not red, but purple or gold, as blood rushed in to help repair damaged cells. Sometimes, in both biology and culture, things just *are* – idiosyncratic products of happenstance.

Marc and I turn the corner to find Maggie and Freya up off the trail, beneath the trees.

"Mom, come look!"

She's found a big *Suillus* mushroom. As I clamber up toward Maggie, my head is full of fungal facts. But when I reach her, squatting beside its velvety umber brown cap, I understand that this moment is not about what I know but what Maggie can share. For once, I keep my mouth shut, listening to her talk, trusting there will be time enough later for her to learn the difference between gilled and pored mushrooms, to marvel at the fungal interconnectedness that underwrites all forests, to wonder at the cause – lots of rain or little frost – that has resulted in such abundant *Suillus* this year.

Autumn and childhood: coloured or not, both are ephemeral. Both change before our very eyes. Right now, what's most interesting to me isn't Maggie's loss of colour (few biological pigments, after all, remain constant), but the potential legacy of the colours she once wore. As Maggie enters the full grip of her teenage years – a developmental stage unknown to the world when the trees around us first germinated – I'm likely to worry more and understand less. Squatting beside Maggie, in a forest dripping gold, I take faith in the ability of ephemeral, cast-aside colours – like carmine dogwood or gold aspen leaves – to continue to *do*. I know that, come next spring, as our oddly canted Earth dips its northern latitudes toward the sun, the chewers and the eaters and the decomposers will rise up. Fungal hyphae will begin to spread. Bacteria will proliferate; invertebrates will tunnel. It will take several years, but the fallen colours of today will fuel other lives before their molecules are fully erased.

Today, what I know with a nonephemeral certainty is that if Maggie finds her way back to red – lipstick or otherwise – in her teenage years, I will bite my tongue. I will refrain from any more *mom as biologist* comments. If my daughter comes home emblazoned with tattoos, like so many students today, I will think of her tattoos not as stains but as colours letting her *do*. And I will encourage myself to live vicariously through all that she does.

What I can't know today, but will by next Easter, is how cohabiting with a teenager thins the line between vicarious and first-hand living. Little do I anticipate the day next spring, when I will, after a pronounced negotiation, follow Maggie into a piercing studio in downtown Kamloops. Nor can I imagine how, after a stressful half-hour, during which Maggie will volunteer to go first, we will both emerge with pierced noses. Hers, with a glittering, but colourless, stud; mine, with a stud coloured as close to quinacridone gold as I could find. Red, I will think, might be just around the corner.

14. Winter's Bottleneck

QUORRK...QUORRK.

Abrupt cries puncture my panic. Tangible and local, the calls halt me in mid-escape. In the echoing silence that follows, my body reverberates with adrenaline, but the cries above have distracted me enough for rational thought to return. I look up.

There, against a winter-blue sky, ravens are congressing, flying in on outstretched fingertips, gregarious and vocal. Some descend to gnarled tree limbs, adding rolling garbles and intermittent *tok, tok*s to the conversation; others fly on. As my heart rate slows, I'm a little embarrassed. I thought I was beyond this. But there's no denying my clamouring leg muscles, the fear-fuelled vibrations still echoing through my body.

On the top of a small knoll, clear sightlines in all directions, I shrug off my pack and pull out my Thermos of tea and sit pad. I'm back at Botany Pond – this time in search of winter. Not the dreary winter that piles up, brown and gritty, on sidewalks in town, nor the anxious winter I remember from my childhood. I want exhilarated winter. The one I've read about in books; the one where children glide down slopes and slip across frozen water. Yet, instead of drawing, I've been panicking, flailing really, through snow.

Now, as my heart continues to calm, I stamp out a flattened area with my skis, stick my poles upright in the snow, blow air into my sit pad, unclick my boots and sit down, my face turned to the sun. I don't have long; it's only 3:30 in the afternoon, but the sun is already close to the western horizon. Atop my sit pad, I open my Thermos and breathe deep.

IN THE NORTHERN LANDSCAPES I've always called home, the season of limited light, nearly by definition, is the season of scarcity. For green plants that eat the sun, less light means less food. Even worse, the tilt of our planet means that each winter day is warmed by fewer hours in the sun's warmth, cooled by more hours in our globe's shadow. Plants may be rooted in place, but when winter allows polar temperatures to creep down from the north, their world transforms from green to white.

The capacity of snow to alter a landscape is so powerful that many biologists believe winter begins only when 20 centimetres of snow has fallen.[1] Yet, in the Thompson Valley, such snow is no longer guaranteed. Last year, we celebrated Christmas with green lawns. But this December, snow fell heavy and hard, bending conifer saplings into snow dragons, magicking cut stumps into madonnas and hobgoblins.

In the face of such transformation, inhabitants must either flee or adapt – both are costly. Sandhill cranes and warblers have fled south from this grassland months ago; bears and badgers are dug into dens. With their rootedness, the options for Botany Pond's plants are more constrained. Most have retreated underground, where they will overwinter as seeds, dormant buds, rhizomes or bulbs. Those whose stems stretch above the snow have no choice but to resist. Either way, there are no guarantees. Whether through dormancy or death, winter hushes a landscape.

IT'S IMPOSSIBLE NOT TO NOTICE. Several hours earlier, when I clicked into my skis and started along the upper trail into Botany Pond, the silence was palpable. The loudest sounds were my own breathing and the *squeek*, *squoosh* of my boots. Even the view – a visual complexity I normally relish – felt muted. Some texture remained – poky Douglas fir, mounded aspen, gnarled old tree bark – but colour, especially at a distance, simplified into a trichromatic palette: snow white, dull dark and shadow blue.

Winter, I soon realized, alters both a landscape's ecology and the attention we give it. Skiing in, I was swept up in the drama captured in the snow's canvas. The abrupt puncture of deer tracks atop a snowmobile's imprint. The impressions of snowshoe hare that floated alongside. A hard-packed highway of red squirrel prints leading to a mounded cache, enveloped within the skirt of an upturned tree. Miniscule, paired impressions of a mouse's hind feet. And then, just beyond the forest edge – a weasel track. Impressions in the snow that doubled back on themselves, abrupt and interrupted. Two animals, or just one? I wasn't sure. Hunting? Or just playing? Again, I couldn't be certain. But, in front of me, these erratic prints were a rare glimpse into the lives I rarely consider in the full thrust of summer's green.

The moment I stepped off the snowmobile trail – away from the deer tracks, intent on making my way down toward Botany Pond – the contingency of snow made itself known. One step away from the track, and I sank more than a foot through powder before the snow could hold my weight. No wonder the deer were so loyal to the hard-packed snowmobile track. Some animals, like the snowshoe hare and the lynx, have disproportionately large, furry feet that act as built-in snowshoes. Deer, in comparison, sink with each step and congregate beneath evergreen trees where the snowpack is thinner. Without skis, I would have been post-holing through hip-deep snow. I was glad I'd left my dog Shasta, with her 11-year-old, stiffening joints, home.

IF SNOW TRANSFORMS LANDSCAPES, each new layer of snow is transformed by the land and weather into which it falls. Even in winter, heat leaks from Earth's core, warming the bottom layers of a snowbank and creating a temperature gradient from snowpack bottom to top. Parts of the snowpack will evaporate from crystal into gas that then diffuses upward. Other parts will melt into liquid that falls earthward. On some days, these interacting gradients will work to cohere billions of snowflakes into a solid mass; on other days, evaporation will form a layer of brittle, loosely arranged depth hoar at the very base of the snowpack.

Changeable or not, the importance of snow in winter's ecology is hard to overestimate. Each new layer of snow works both as blanket and mirror, capturing radiant heat released from Earth, and bouncing the light of our sun away. I know that, somewhere in this landscape, grouse hunch in the upper levels of the snowpack, coming out to feed on catkins, leaving behind their tubular pellets of scat. Deeper, near the interface between ground and snow, where the isolated crystals of depth hoar make travel easier, small rodents like mice and voles flee hunting weasels.

TODAY, THE SNOW also tells my story. From my perch on this small knoll, I can retrace the twinned lines left by my skis. Headed downhill toward the aspen stands just above the pond, I'd travelled, for the first time, at more than a snail's pace. En

Deep snow is winter's prism. Pigments simplify, textures soften, paths alter. In this place trees bear crystals of light, bird nests pile into wobbly snowcaps and mysterious weasels leave abundant sign.

This is winter's game. Feel the quiet: migration, dormancy and death winnow exuberance into endurance. For those that remain, one species' risk is another's opportunity. Tomorrow's hope depends upon today's bitterness. Below ground, seeds chill, above ground, moisture collects. Rest easy: come spring, rivers will fill, flowers will bloom and weasels will do their crazy dance through tall grass.

Winter's Press, January in the Grasslands

mouse

snowshoe hare

bird nest in water birch

In the impressionable canvas of freshly-fallen snow drama can run quiet or wild. Prints of miniscule mouse barely indent. Snowshoe hare float this way, then that.

Erratic weasel prints explode across the ridgetop, their fierce energy writing complicated story in snow.

weasel

aspen draped in hoar frost

view looking east across Lac Du Bois

Snow falls heavy and hard, transforming saplings into snowdragons lumbering across the trail,

and magicks cut stumps into madonnas and hobgoblins.

Studio Illustration: Winter's Press, January in the Grasslands

route, I'd imagined my body weight compacting the layers beneath me, even as my skis gained speed. In the white-on-white surface, it was difficult to distinguish the bump of a log from a wind-piled fluff of snow. I'd flexed my quadriceps, bent my knees and forced the edges of my skis into a careful snowplow. My skis glissaded, buried in the uppermost layer of powder. By the time the slope levelled out, and I had come even with an aspen stand, a wide smile ranged across my face. This, I thought, was surely one of winter's exhilarations – gliding suspended between earth and air, through a crystalline matrix of frozen water.

Closer to the aspen stand, I found another. Aspen branches and twigs – each draped in sparkling dendrites of hoar frost. I'd stood there, entranced, imagining the mysterious calculus of temperature and diffusion gradients that could allow water vapour to condense directly into ice on these trees. In front of me, draped in ice, aspen branches bent and looped into extraordinary coils.

External ice is one thing – the weight of ice can break tree branches – but the real danger for plants lies with internal ice. All living cells slosh with the water necessary for the chemical reactions of life. But if this water cools into ice, not only does it become unavailable for plant use, but its tiny needles can pierce cell membranes. Animals prevent this water from freezing by retreating to warmer places beneath snow or ground. But trees have no place to go.

Instead, during the months leading up to winter, trees undergo a suite of carefully orchestrated changes in cell chemistry, including, in some, the production of molecules that act as antifreeze. Between species, the details differ, but the intent is the same: to maintain cellular water in its liquid form, even if this allows ice to invade crevices found between cells. These plants wear snow inside, isolated and contained.

Plant or animal. Inside or outside. Aided by natural selection, or human manufacture. Whether it is a down jacket, or an adept chemistry, we endure winter through bodily transformation, our capacity to survive winter shaped by the innovations of those who came before. Yet part of winter's novelty lies in its capacity to redefine the world we take for granted. Like all seasons, winter encompasses both possibility and peril. This was the lesson I understood, not intellectually but viscerally, deep in muscle and bone, the moment I skied across tracks unlike any I had seen so far.

These were not the tracks of individual feet, but rather the imprint of an entire body – a body that was moving at speed, with the energy and strength to bound through hip-deep snow. Each imprint was nearly the length of my ski pole, separated from the next by nearly twice that distance. Clearly, the animal that had left these tracks was far larger than a weasel. As I sorted through the possibilities, I'd felt my entire body go alert. This was the trace of a carnivore – maybe just a coyote, but the distance between bounds seemed too large. Lynx? Or wolf? Both, I knew, had been caught on motion-activated cameras nearby. Both could be hidden from my view behind the aspen. Without conscious thought, I tore myself away from the ice-covered trees, and skied as fast as I could toward open grassland. I wanted my dog. Without Shasta's nose and behaviour, I had no idea how recent these tracks were. I felt like prey.

In a single moment, winter's exhilaration slid past anxiety into terror. Yet, even as I fled for the comfort of unobstructed sightlines, my fear felt familiar, something I'd been expecting. That's when the raven's call, like a slap in the face, dropped me back into my body.

FOR MORE THAN FOUR DECADES, my life has been marked by winter's imprint. Growing up, I'm sure there were exhilarated moments as I played in the snow, but I don't remember them. Instead, in my childhood, winter was the anxious consideration of our woodpile's height against our woodstove's appetite. Winter was fingers of ice spreading across our windows' inner surface. Winter was shivering in the canopy-covered back of our small pickup, while my mother lay underneath, putting on the tire chains necessary to drive uphill through the snow drifts on Irish Creek Road. Winter was the dread that our water pipes would freeze, and we would be relegated once again to filling old milk cans from the water spigot in front of the church. Winter was trudging through ankle-biting, cold-hardened snow.

It wasn't until I moved into my university dorm that I learned how central heating could defang winter's bitterness. That first winter away from home, the thermostat in my dorm room was nothing short of miraculous. Later, as part of my graduate coursework, I would spend two weeks in a cabin in Maine, studying winter ecology. By then I had skis, good mitts and hat, thick Capilene underwear, a down jacket and Sorel boots, but I was never really convinced. Winter still meant trying to squeeze extra cash out of my grad student stipend to pay the oil bill, and finding myself where my mother once lay – in the deep snow, under a truck, fighting with icy chains. My love affair with plants began in the long, sweet suck of summer, when the world is green and fecund.

I was not wrong to fear winter. Although humans have been living with winter for millennia, our biology is still largely that of tropical or subtropical apes. Millennia ago, when our ancestors moved into temperate regions, they survived by wearing the fur or feathers of other species, sheltering in caves and corralling fire. If these cultural innovations helped mediate the climatic extremes of winter, they still required extra time and energy to acquire. Resources that any one individual, any one family, any one community, may or may not have. Climatic extremes, I realize, are rarely without inequity.

Today, in the landscape of my adulthood, buoyed by the possession of a regular paycheque, a house plumbed for gas, and a car that starts when I need it, I can clamber over winter's extremes. Today has been more than just a field trip in search of winter; in its own way, it's been a recognition of the tide turn in my own life. Today, I can imagine the possibility of an exhilarated winter.

It took three tries to get here. Last week, I came with Marc and Maggie and Shasta, hoping for sun over snow, and found only the murky inversion that regularly clouds the South Thompson in winter. Yesterday the inversion cleared, but I was too late, and the snow was too deep for Shasta's elderly bones. This morning, I planned carefully. Timing was critical. Too early, and my trip would have been an exercise in cold resistance. Too late, and the pull-offs where I could park my van without getting stuck in the snow would already be occupied. I left Shasta at home, resisting her doleful glances. I double-checked that I had the necessary gear – skis, mitts, multiple layers, Thermos of tea. And it worked. Gliding suspended through snow. Fleeing from predators. Winter's exhilaration beats most strongly, I think, when its joy is sensed in full light of its risks.

Now, as the ravens still call, my fear has nearly ebbed, and there are a few moments of sun left. I tilt my head back and push sound out from deep in my larynx, wanting to acknowledge their gift. I see one big black shape angle its head down, peering at my vocal intrusion, even as it flies steadily on.

As I finish my tea, I understand. Snapshots from a single season never tell the whole story. Turn down the thermostat; let rain fall as snow, and the known becomes unknown. I came looking for winter, but I needed to remember that a dark belly of conjoined need and opportunity always rides beneath winter's white coat. Winter is the prism through which trees bear crystals of light in place of leaves, and mysterious weasels reveal their lean presence. There is even an odd comfort in the harshness. For many plants, summer depends upon winter's transformation. The wildflowers that fill this grassland in June won't germinate without a sequence of cold days; Botany Pond fills only via snowmelt.

In this part of North America, winter is the bottleneck that defines much of the rest of the year – biologically, culturally, financially. Winter always balances between resilience and vulnerability, between plenitude and disaster. But all of us – aspen or weasel, human or wolf – survive seasonal extremes best when they are dependable. When what has worked before will work again. Today, the real danger of winter lies not with its extreme, but with its *increasing* unpredictability.

It's time to go. The sun has fallen past the horizon, and the cold is advancing. I pack and click into my skies. I can rest easy: with nearly a metre of snow on the ground, the ecology of winter is doing what the land needs. For now, I follow my trail back, reassured that, this summer, rivers will fill, flowers will bloom, and out of my knowing, weasels will do their crazy dance through tall grasses.

15. Collecting the Grip

ON A SUNDAY AFTERNOON in mid-March, I open my truck door and step into the high, thin light of a spring afternoon in the upper Lac du Bois grasslands. I am, once more, headed into Botany Pond. Around me, the hills are brown and muted from the weight of winter. I'm layering up – hiking boots, down jacket, windbreaker, knit cap, gloves, binoculars – when my eyes drift skyward. There, perched on a whip-thin aspen, a raptor stares down at me. I reach back into the truck, one hand scrabbling for field journal and pencil, even as I bring binoculars to eyes. Knowing that my time with this bird will be shorter than I'd like, I make notes as fast as I can, willing my body to be a conduit: eye-brain-hand-page. The bird faces toward me: lighter than a red-tailed hawk, clear white belly, speckled breast band, dark tail, dark mask on its face circling down to breast. As its head swivels away and then back, a hooked bill (*maybe with black at the tip?*) flashes past. Then it's aloft. Broad wings, clearly a buteo, and not a falcon. The bird cants through sky, a white band slicing across its tail, until the nearby hill swallows it whole. My pencil stops. Silence holds.

And then starlings across the road chatter the day back into sound. Time regains its normal footing. Crossing the gravel road, I think about the moment I just shared with the hawk. I am different after having felt a raptor's glare skittering across and then dismissing my form. But even now, years after I first learned to hunger for such moments, I can't predict their occurrence. Other than putting myself in place, field journal in hand, I can do little to facilitate them. What I do know is that, in the Anthropocene, collecting these moments feels like a kind of hope.

Collections are often defined as "groups of things": plant specimens, plastic figurines, poems. Beyond this, the limits of collecting, and by inference, of collections, are surprisingly contentious. Some argue that collecting is a basic part

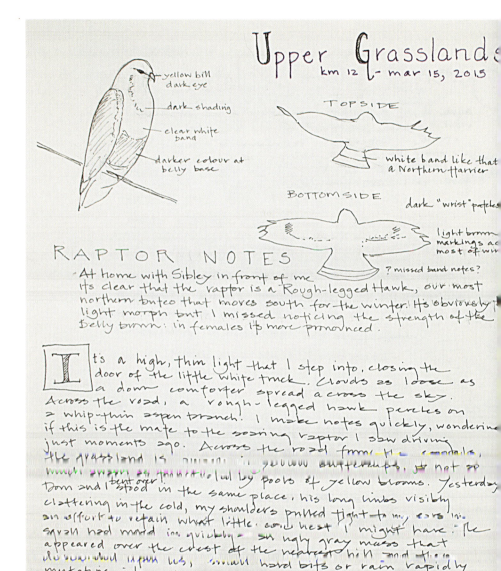

Upper Grasslands

- yellow bill
- dark eye
- dark shading
- clear white band
- darker colour at belly base

TOPSIDE

white band like that a Northern Harrier

BOTTOM SIDE

dark "wrist" patches

light brown markings ac most of wi

? missed band notes?

RAPTOR NOTES

At home with Sibley in front of me it's clear that the raptor is a Rough-legged Hawk, our most northern buteo that moves south for the winter! It's obviously light morph but I missed noticing the strength of the belly brown: in females it's more pronounced.

It's a high, thin light that I step into, closing the door of the little white truck. Clouds as loose as a down comforter spread across the sky. Across the road, a rough-legged hawk perches on a whip-thin aspen branch! I make notes quickly, wondering if this is the mate to the soaring raptor I saw driving just moments ago. Across the road from the cattails, the grassland is

Dom and I stood in the same place, his long limbs visibly clattering in the cold, my shoulders pulled tight to my ears in an effort to retain what little

appeared over the crest of the nearest hill and then

mutating into wet snowflakes that soaked the ground we were sitting/kneeling on.

Today in the company of sagebrush buttercups and Indian potatoes / spring beauty, I can be slow enough to hear the presence of other species.

Upper Grasslands
— p h e n o l o g y —

mar 15
- Ranunculus glaberrimus var/ssp ellipticus (f)
- Claytonia lanceolata (f)
- Zygaden venenosus - lvs just emerging, 1 in above ground
- Poaceae, lvs emerging (2 in higher) more abundant where grazed
- Flicker calling
- ⊠ Rough-legged hawk, pair flying: one perched
- Aspen buds "in fur"
- Bashful fritillary (F. pudica) in bloom

ASPEN CATKIN

CLAYTONIA

CLAYTONIA, FACE VIEW

delphinium leaves

BASHFUL FRITILLARY

Castilleja

- Robin calling
- Northern shrike
- Starlings
- Ice still on N·facing slopes
- Mtn bluebird pair
- Lomatium macrocarpum lvs.

of the human experience, a remnant of the foraging instinct found in all animals, and includes any gathering of objects.[1] Collections, these experts argue, can be amassed without direct intent, with a significance, if any, understood long after their initial collection.

Others insist that collecting includes only the deliberate gathering of "things of subjective value."[2] Among those collections made with intent, some may be temporary, destined for one season; many more aspire for permanency. Nearly everyone agrees that diverse motivations prompt our collecting: a way to pass time, a quest for wealth or knowledge, a salve for a childhood hurt, a concern for the future.

Yesterday, a research student, Dominic, and I spent the morning counting the number of sagebrush buttercup flowers in plots on the hills just south of Botany Pond. Most we left in place, a few we collected. The data set we tabulated in a Rite-in-the-Rain field book is good; it'll add to our understanding of this species, the first flower to bloom in these grasslands. Yet, even as we shook and shivered beneath passing snow squalls, I understood we weren't collecting merely numbers. Last autumn, during unseasonably warm temperatures, I found sagebrush buttercups blooming in November. A whole hillside: five months too early, a full season out of sync, a yellow flag of the Anthropocene's growing unpredictability. After a winter of worry, yesterday's data set reassured me; today I woke wanting still more comfort.

Collecting is rarely without risk. To both the collector and the collected. Sagebrush buttercup's Latin name, *Ranunculus glaberrimus*, is based on a specimen collected near Kettle Falls on the Columbia River, by the Scottish plant collector, David Douglas, in 1826. A collector who inexplicably lost his life in Hawaii, while travelling home from his third trip to North America. In the grasslands throughout the intermontane west, sagebrush buttercups root in place, entangled with soil bacteria, pollinators and seed dispersers. In the herbarium of Kew Gardens just outside London, England, the buttercup life Douglas collected nearly 200 years ago – its dried carcass, flattened and glued onto an individual sheet of paper – stacks among other specimens in a metal case. Intact but inert.

It's hard for me to imagine the determined focus of botany's most famous collectors, yet many of us work long hours to collect "groups of things." Author Yuval Noah Harari writes, "Over the course of his or her life, a typical member of a modern affluent society will own several million artifacts – from cars and houses to disposable napkins and milk cartons."[3]

If we all collect, what collections do we want to most shape our lives?

AS A PRACTICE, the science of plant collecting co-evolved with the European project to name and claim the world. In doing so, its techniques transformed rooted beings into specimens that could be shipped across the world. For nearly six centuries, the tools and techniques of plant collections have remained largely consistent: hand lens, digging tool, plant press, field journal, glue, mounting paper. David Douglas – whose botanical travels brought him through the Thompson Valley during the spring of 1833, just over 20 years after the Tk'emlúps te Secwépemc first hosted European traders in search of furs – faced greater challenges than I do in sourcing good paper (and keeping it dry), but we could have easily shared our gear across the centuries. And, for both of us, there's no doubt that learning to collect plants transformed our lives. Collecting gave Douglas, the son of a Scottish stonemason, reason to travel the world, to mix with the scientific elite of his time. Collecting gave me a profession, a reason to return home.

Like Douglas, I learned the craft of plant collecting from my elders, most often in the field. Plant collecting, done as much with the body as with the mind, is a tradition handed down from one generation to the next, entangling people and place with specimen and convention. Maybe that's the reason the lessons of my teachers – along with their hopes and habits – return to me most frequently in the field. None more than the most accomplished collector of my teachers, the late Dr. Wilf Schofield. Like most botanists, Wilf labelled each specimen he collected with a sequential number. Recorded in his field book, this number was the numeric code that forever links specimen to place – or at least to the abstraction of place Wilf recorded in his field book.

Referred to affectionately as a "garbage-bagger," Wilf was renowned for both the number and sheer mass of his collections. Wilf, along with his students, built the collection of bryophytes – the small, nonvascular members of the plant world – at UBC into one containing more than 250,000 specimens. Collections never end (that is, after all, a big part of their lure), but collectors do. When Wilf died in 2008, at the age of 81, his own collection number had reached a staggering 128,619. He assigned this number to a moss he'd gathered in the last summer of his life on an island in the Aleutian archipelago – just days after the island's volcano had erupted.

Numbered or not, every collection searches for something. In 1753, the Swedish botanist, Carolus Linnaeus, published what he believed was a complete list of the world's plants in his book, *Species Plantarum*. In consistently referring to each plant species with a two-part Latin name (think *Pseudotsuga menziesii, Ranunculus glaberrimus*), Linnaeus inadvertently created the linguistic infrastructure to support a globalized botany. But on this historians are clear: the intent of Linnaeus's collection was to *acquire*, but not to *colonize*. During Linnaeus's lifetime, Europeans, still largely confined to their continent, had already developed strong appetites for plants grown elsewhere: coffee from Africa, tea and sugar from India, oranges and opium from China, cinnamon from Sri Lanka, tobacco from the Americas. Linnaeus wanted to find the plants – either species native to Sweden, or exotics that could be cultivated in Sweden – to serve as substitutes for the plants his country was going into debt to acquire. Collecting, for Linnaeus, was not just a method to catalogue the world's botanical diversity but a mechanism for Swedes to eat, drink and smoke local.[4]

Yet all collections risk unforeseen consequences. Linnaeus may not have intended to unleash a colonial empire, but botany has always been as much a business as it is a science. With their continent home to only 12,000 of the world's nearly 400,000 plant species, every trip abroad allowed European botanists to bring home novel plants. The European appetite for exotic plants quickly spread beyond what could be consumed, and by the early 19th century, plant collectors like David Douglas were prospecting across the continents for the plants that could fill European garden beds and their cabinets of curiosities.

Loss haunts every collection – some more than others. Although the Secwépemc People relied on these bright yellow buttercups as harbingers of spring,[5] Europeans had little use for this plant outside their science or gardens. Unlike tobacco, sugar or rice, this plant was never collected into those brutal fields of forced labour called plantations. The production of its fruits never relied upon radically simplified ecologies, the dispossession of millions of Indigenous plants and peoples, or the displacement of millions of enslaved lives from one continent to another. Unlike with potatoes, corn or wheat, the shuffling of sagebrush buttercup plants across the globe never produced the caloric or monetary surpluses that helped fuel European population growth and the establishment of colonies across the world.

If collecting some plants helped build the infrastructure of the Anthropocene, other collections have tracked its impact. Linnaeus didn't create the world's first herbarium, but he thought every botanist should make one. And, for the last six centuries, botanists have. I think of those hands – some like those of Charles Darwin, or David Douglas, or Alexander von Humboldt, whose names still echo across the centuries; the many more whose names are rarely uttered outside herbaria – that selected, dug, dried, spread, glued, annotated, and continue to care for the 390 million specimens collected in the world's herbaria.[6] One specimen for every 1.3 square kilometres of our Earth's surface. I think of those green lives – stolen from home, isolated from community, suspended into specimen – that still have enormous lessons to teach us about our world. Climate change, pollution, species extinction. Tied by a collection number to a particular time and place, these lives – once responding, now preserved in the world's herbaria – track the change we humans have wrought on the world. Looking back, botanists now understand, may well help navigate the future.[7]

MAYBE IT'S JUST MY BIAS, but I can't help but think that the wisdom of any plant collection extends beyond the lessons of its specimens. Collecting, no matter the intent, necessitates that we venture into the world of living, breathing, green things. Walking past the buttercups that Dominic and I sampled yesterday, I know I will spend the rest of the day hunting for more moments like the one I shared with the hawk. If I'm lucky, I'll walk back to my truck with a few pages of sketches and notes in my field journal.

Yet others have already recorded much of what I will observe in Botany Pond. As with most places in our world, satellites I can't even see have imaged the botany of this grassland. I never got the chance to bring Wilf to Botany Pond, but one of his former students had already inventoried its plants before I arrived. In the Anthropocene, other devices – trail cameras, iButtons, geolocators, PIT tags, RFID readers – collect yet more data nearly year-round. As one of BC's few protected grasslands, and one that is located less than 20 kilometres from the university where I teach, this place has been poked and prodded by scientists of all stripes. Quite literally, I could suffocate under the weight of paper printed with facts *others* have learned about this place.

Even though I've field-journalled more often at Botany Pond than any other place I can name, I wonder how much of its story I've gathered second-hand.

Does my collection of plant stories from this place even matter? In his biophilia hypothesis, E.O. Wilson argues that humans are biologically predisposed to pay attention to the natural world,[8] but I think up until now I've never stayed in place long enough for *attention* to transform into *relationship*.

Maybe that's what makes my collecting at Botany Pond, field journal in hand, so compelling to me. This is learning unplugged, free-range, year after year, across the lives of plants rooted in place. No confining disciplines, no guiding commentary, no guaranteed outcomes. Today in Botany Pond, hawks soar, and sagebrush buttercups bloom. Last summer, not far from here, I sat drawing beneath a Douglas fir tree and was perplexed by the repeated journeys of a red-breasted nuthatch from canopy to leaf litter to trunk. I diagrammed the behaviour in my field journal – bird, tree, ground – until it finally dawned on me what I was missing. Seeds! Together, tree and bird taught me about their own lives: the slow quiet of an August day that accompanies seed fall, the avian attention that collects in anticipation of winter's scarcity. Together, tree and bird became less individual species to track, more members of community, maybe even kin.

Today, I've circled out across the rolling hills, and then back toward a narrow string of aspen trees. My truck's not far. E.B. White – a man immersed enough in the living world to convincingly write the lessons a spider and a barnyard pig might teach one another – once said that a blank sheet of paper was "more promising than a silver cloud and prettier than a red wagon."[9] I look down at my page. Its promise has turned into a list of springtime happenings: first flowers beginning to bloom, first bluebirds fly catching from the lower branches of an aspen tree, last ice still lodged on north-facing slopes. This collection is open-ended: one that builds on observations made in years past, one that will call me back, again and again.

TODAY, I DON'T COLLECT as much as I should. Not scientifically. But many collectors apprentice first with "things" different than those that will form the dominant collection of their lives. Since beginning my project to re-story my relationship with plants, my scientific collecting number has been stalled in the three-digit range. It will be unlikely, I think, to ever exceed four digits. But I think we collect who we are. All day, every day.

If we're lucky, our collecting requires the talents we have available. If we're really lucky, the collections we make require our direct experience – informed, but not swamped, by others – of the world around us. And, if we're exceptionally lucky, our collections challenge our understanding of not just what we collect, but *how* we see the world. Nothing allies us more powerfully *with* the world than the lessons we learn *from* it.

Europeans like David Douglas and Charles Darwin may have set out to explore the world, but their field records reveal that the relationship between the collector and the collected is never one-sided. Reread by historians, the notes and sketches, even paintings, made by these Europeans travellers reflect that, while these men may have left home with definite beliefs about the way the world worked – its imperialism and patriarchy, even its origins and aesthetics – their minds were "in the process of becoming, not fixed and certain."[10] Certainly, few question the impact of Darwin's natural history (including collecting more than 1,000 plant specimens while he sailed on the HMS *Beagle*) on the man himself. Like any skilled or creative practice, collecting transforms our understanding of who we are through a mutual entanglement of world and will.[11]

BACK AT MY TRUCK, I open the door and sling my pack and field journal onto the bench seat. Driving home, hills roll by on either side. Around one corner, I spot a hawk, perhaps the same one I saw before, spiralling upward. I wonder what it's hunting. I glance over at my field journal. It's only a week old, but already its pages are filling. In a couple of months, I'll add it to the stack that sits on my bookshelf. Several years ago, when I set out to re-story our *experience* with plants, I thought I understood the risks of my new collection: the unfamiliar terrain of art galleries, the inchoate terror of reading my words aloud to strangers.

But, in retrospect, it's clear how little I understood the potency of plants in place. The way that plants in place, rich in meaning and memory, can serve as a refracting mirror for our own history. Little did I understand how the ecology of grasslands, the shape and colour of tree leaves, the layered possibility of a peatland, the heavy weight of snow, would entangle with the hard bits of my own family history, my own mothering today. Perhaps that's the real power of collections: the way they begin to answer the questions we have yet to find the courage to ask.

I know that the intent of my collecting has diverged from that of my teachers. In multiple herbaria, more than half of the type specimens – the preserved specimens that botanists use to define entire species – have been collected by just 2 per cent of their collectors,[12] the heavy hitters like Wilf. Collecting for stories means I now collect like a magpie, and little of it can be deposited in a herbarium cabinet. But no collection has influenced me more than the field journals and paintings I put together as my first art show. The January night the show opened in Kamloops's historic Courthouse Gallery, all its images – botanical or otherwise – celebrated the place I call home. All its images – big or small – were encased in the shadow boxes, dovetailed and elegant, built by my husband from the warm, clear grain of bigleaf maple. More than five years later, I still meet strangers who know too much about me from this exhibit.

Maybe it's not *what* we collect, but what our collection *does* that matters. In science, "collect" means to gather systematically. In the Catholic and Anglican churches, a *collect* is a prayer offered up at a certain time. I wonder if both aren't a kind of listening, a way of attending. Maybe the best collections are those that make us open and porous to the world. And maybe, in the new, unpredictable world of the Anthropocene, learning not just to collect but to listen to and learn from plants in their places – whether through specimen or line, worry or joy – is what I've most needed to do.

I will always treasure my memories of Wilf in the field – bright yellow rain pants and faded green slicker cloaking his tall frame, worn canvas bag bulging with bryophytes slung over one shoulder, gray hair pushed awry by wind and rain, numbering up prayers, one by one, for the plants he loved. I will remember Wilf's collection numbers as marking both individual specimens and moments of hope. I think Wilf, and maybe both David Douglas and Carolus Linnaeus before him, listened best when depositing herbarium specimens. I know I listen best when the world grips me as tightly as a hawk clutches a freshly killed vole.

A sorted collection is no miscellany; it is a physical manifestation of what we know to be true and important. Today, driving the last curves of the gravel road home from Botany Pond, I understand that the collection that has allowed me to make sense of my own life, that has reimagined my own botany, is the collection of field journals standing quiet and inert on my bookshelf.

No wonder the current volume comes with me everywhere.

PART III

A Canopy
of
Care

CLEARWATER VALLEY

BOTANY POND HEFFLEY LAKE

KAMLOOPS

Pacific
Ocean

FRASER RIVER

COLUMBIA RIVER

KOOTENAY RIVER

MISSOURI RIVER

COLUMBIA RIVER

MILL CREEK

N

Location Map for a Canopy of Care

16. Carrying Capacity

FROM THE MOMENT we hop out of my van, it's clear. I've prepared for this ski atop the frozen surface of Heffley Lake with my usual pack-mule mentality. My friend, Elizabeth, in comparison, is whippet-like, lean and trim, with just one layer of outer gear and her skis. I'm so impressed that I'm tempted to ask if she has anything in her pockets, but that's too nosy – even for me. What I do know is that I'll be embarrassed to carry my normal burden beside someone who seems to need so little. Behind us, nuthatches *eeek, eeek, eeek*, and a raven *carooks* in the far distance. Below us, the lake's frozen expanse is cupped in the linear hollow of this high valley.

Opening the back of my van, I regard the messy pile of gear. What to carry? I ask this question so often, I'm sick of it. Anytime I go into the field, my pack bulges with a shifting mix of tools: illustrated field journal, field guides, hand lens, sampling quadrats, datasheets, pens, pencils, paintbox, waterbrush, camera. It's a rare trip when I use it all, but it's always difficult to leave anything behind.

Ecologists use the term *carrying capacity*, meaning the number of individuals a place can support without environmental degradation. Each of us, I think, has our own carrying capacity for the number of objects we can support without mental degradation. Marc travels light, no matter where we go. He's happiest if he can walk with hands and back unencumbered, the objects in his pocket more than meeting his carrying capacity. Standing at the van, my guess is that Elizabeth's carrying capacity runs similar to Marc's. Maggie – well, she'd bring the kitchen sink if she could. And I fit somewhere in between, carrying more than I should, but more restrained than my daughter.

Of course, *what* we carry reflects more than just personality: it tangles with the legacy of the past, the intent of the future. My dog Shasta's quivering nose, the instincts that drove her to bound disobediently after deer, were traits she inherited from her wolf ancestors. Her floppy ears, her tendency to cock her head to one side when she was interested in what I was saying, signalled her species' ongoing relationship with mine. Certainly, Shasta was keenly aware of the promise implied the moment I stuffed her leash in my pack.

This morning, however, I keep thinking about the reciprocal relationship between *caring* and *carrying*. If our *caring* dictates what we carry, can what we *carry* influence our care? Nine months of pregnancy with Maggie transformed an ambivalent notion of motherhood as something I probably shouldn't miss into the piercing emotion I felt for the squalling infant who arrived in my arms, skin red and blotched from her travels. By the time I found Shasta, gangly and half-grown in a cage at the pound, she was already too big to carry. Yet, for the last 12 years, first along with our older dog, and then just by herself, Shasta's companionship has been part of what I've carried into the field. But I can rely on it no longer: Shasta died four days ago.

WE'RE DOWN ON THE LAKE. Elizabeth turns to the left, following the tracks she made earlier this week. Our skis swish beneath us; cold pulls at exposed skin. On the horizon, snow-draped conifers pitch themselves across the rounded topography of Embleton Mountain. As the lakeshore unfolds, frozen point by secluded bay, the conversation between Elizabeth and me rambles. This place is populated by Elizabeth's stories, not mine. More than 30 years ago, Elizabeth and her political scientist-turned-carpenter husband, Monty, settled just uphill from the lake: building a house, raising three kids, and tending a passel of pigs and sheep. Monty, Elizabeth says, has either built or worked on most of the homes we ski past.

This lake, perched near the upper limit of Douglas fir forest, is 35 minutes and one ecosystem away from my home in Kamloops. Today, it's my refuge — a landscape unburdened with memories of my dog.

Carrying capacities are, by definition, limits. Problems arise with too many mouths or tools. Even today, when I abandon my backpack in favour of my compact, orange field bag, I can't find the topo map in any of its pockets when I want to orient myself. Finally, in frustration, I shake my field journal upside down, and the map's folded page flutters out. Sometimes less is more; sometimes having the right tool matters. I say different things with my silky-smooth Pilot G-TEC-C4 pen than I do with my Micron PIGMA. And my new Uniball SIGNO's scratchy nib provides the perfect tension for drawing but is too slow for writing.

Standing upright, I blow snow crystals off the map and orient it with the landscape's rhythm. The map is never the territory, but right now I value this map's perspective. Looking up, and then down and back again, I translate territory into map – that line of trees is an island in the bay, and that far-off gap is the last of the bays that pinch off, amoeba-like, from the east end of the lake. As Elizabeth swishes away, and I turn to follow, I wonder if, today, my impatience with my tools has less to do with their quantity and more to do with how poorly they compare with the companionship of my dog.

I never thought of Shasta as a *tool*, but there is no doubt her company was a conduit to the world. More than once, when I was immersed in a drawing, her sudden alertness – a raising of head or hackles – warned me to look up. More than once, when I lost my bearings, her nose smelled the right path home. Paleoanthropologist Pat Shipman argues that the ability to *connect* with animals – cultivated first to avoid being eaten by them, then in learning how to hunt them, and finally in learning how domesticate them as "living tools" – was a potent force in human evolution.[1] As our first domesticate, no animal has been more important to us than the dog.

The length of this relationship – some say as long as 33,000 years, others argue only 15,000 years[2] – has been long enough to mark both species. Given our co-existence, it's not surprising that many of the genes related to diet, digestion and disease in dogs and humans have evolved in parallel.[3] Living with us has also changed dogs' morphology – their skulls differ in both size and shape from wolves – and their behaviour. Dogs are particularly adept in reading human social cues, understanding us better than wolves or even chimpanzees,[4] our closest evolutionary relatives.

The consequence for humans has been no less dramatic. Put simply, Pat Shipman believes our connection with animals *is* what makes us human. Or at least she says this connection is deeply implicated in the evolutionary transitions we've made as a species. Our hominid ancestors, she explains, "transformed rock into stone tools and stone tools transformed hominids from bipedal apes that are basically herbivorous (plant-eating) into predators."[5] Predation demanded our deep

attention *to* and understanding *of* those animals we wanted to eat. Language –
another evolutionary transition – was particularly adaptive, as it gave us a way
to both share and store our knowledge of animals. Talking *about* animals, she
says, triggered yet one more evolutionary innovation, in that it led to talking *with*
animals – the symbolic communication that sits at the heart of our relationship
with domesticated animals.

Who first initiated domestication's mutual dependency – wolves or humans –
is still unclear. This relationship might have started when individual wolves
adopted humans as a source of prehistoric table scraps. Or maybe it started when
someone brought home an orphaned litter of wolf pups, raising and then selecting,
over generations, those pups who behaved best as dogs. Either way, it's clear that
when wolves and humans first met in Europe, they forged a mutually beneficial
alliance.[6] Humans profited from dogs' superior noses, fleet limbs, barking alarms.
Dogs gained access to weapons that could be thrown (spears, clubs and atlatls) or
hidden (snares, pits). With dogs at our side, we humans spread to every continent
on Earth, and humans have enabled dogs, now numbering nearly one billion,[7] to
become the most abundant carnivore on Earth.

With dogs and humans, I don't know if it matters who's the tool and who's
the tool user. I'm happy, maybe even honoured, if Shasta thought of me as her
slow-footed ape who supplied her with kibble and tagged along at the end of her
leash. What's important to me is that, if Shasta and I were each other's tools, it was
reciprocal care – mediated through the hormone, oxytocin – that underwrote our
relationship. Sometimes called the parenting hormone, oxytocin flooded through
my body when Maggie was born. When she was a baby, even our looking at each
other – what psychologists call a mutual gaze – could spike oxytocin levels in both
our bodies. The same mechanism cements the bond between dogs and people. In
a recent experiment, oxytocin levels increased when dogs and their owners gazed
at each other, but did not change when the same experiment was replicated with
tamed wolves and their owners.[8]

Is it any wonder, I think, that Marc and I decided to become parents not long
after we began our life with dogs? If caring for dogs primed me to risk motherhood,
then their deaths – first our older dog and now Shasta – have given my daughter
her first grief. I may never have carried Shasta in a sling like I did with Maggie,
but as my skis (yet more tools) slide me along, I know I will carry her absence for
many months.

LIKE MANY RELATIONSHIPS, the one between carrying and caring is complicated. After I return home from my ski with Elizabeth, my muscles comfortably sore (Elizabeth not only carries little, she skis fast), I look up the two words in the dictionary. Given their shared syllable, I always thought they shared a similar origin. I was wrong. "Carry" comes from the Latin word *carrus*, meaning wheeled vehicle, whereas "care" comes from the Old German word *chara*, meaning grief or lament.

Tool, emotion. Dogs teach us that either one can lead to the other. It's not that Shasta's death was unexpected. She was diagnosed with liver cancer several months ago, and when the drugs no longer masked the pain, Marc called the vet. Four days ago, I thought I was ready, but nothing could prepare me to watch Shasta's head slumping into Marc's arms. In the 12 years she spent with us, it was the first and only time she let any of us carry her weight.

SEVERAL DAYS AFTER MY SKI with Elizabeth, I go back to Heffley Lake. This time, I carry my full backpack, including what I need to stay warm sketching outside: a blow-up sit pad, waterproof ground sheet, Thermos of hot tea. My trip is slower, and I don't travel nearly as far as I did with Elizabeth. Alone, Shasta's absence weighs heavier. But if my dog taught me anything, she taught me the value of attending to the world outside my head. When I arrive in a wind-protected cove on the far side of the lake, the fumbles I make with my gear (dropping my camera in the snow, forgetting to sweep the snow off my waterproof sheet before I sit down) feel less frustrating.

Shasta opened the door to the world beyond my senses; my field journal and art supplies cultivate the attention capable of transforming the general into specific, the mundane into momentous. I would gladly trade every pen in my pack for one more day of Shasta's companionship, but, today, something changed the moment I put pen to paper. It's not just that I saw more – patterns in the windswept snow, the play of light across the hills – but I *thought* differently. Master woodworker Peter Korn, in his book *Why We Make Things and Why It Matters*, explains that any creative practice tests both the medium of the practice – wood, paint, sound, language or the human body – and the mental maps of the artist.[9] Settled into my protected perch, I used my gloved hands to sketch two landscape views – first, the interlocking lines of Sawmill Bay that I skied around with Elizabeth, and then the steeper topography of Embleton Mountain, a peak I have yet to climb. On

Illustrated Map: Cupped in a Linear Hollow

this cold but sunny day, my medium is paper and pencil, pen and paint; but what mental map am I challenging?

Beside me, sapsucker holes ring a birch trunk, and alder branches hang low. On the alder, a cluster of appendages twist in sensuous yet disturbing curves around a catkin. Pencil finds paper, and I'm already drawing when I realize what's just happened. Skiing over, a list had grown in my mind: cottonwood, Douglas fir, spruce, paper birch, mountain alder. My experience of these species had been a generalized listing – less of individual lives than of Platonic ideals – that allowed me to remain comfortably distant from nature. But the intimacy demanded by drawing – the deep attention to an individual life, deformed, perhaps besieged by disease – removed all pretense of objectivity. Later, I will read artist John Berger's description of this phenomenon: "the contours you have drawn no longer marking the edge of what you have seen, but the edge of what you have become."[10] Later, I will learn it's a fungus called alder tongue (*Taphrina occidentalis*) that has subverted catkin tissue into these strange extensions. But, in this moment, I care less about mental maps or fungal biology and more about the emotion that just rushed in alongside the lines I drew.

Right here, right now, it's clear. Drawing breeds intimacy; intimacy risks heart. The moment I recognized this alder in front of me as an individual with its own struggles of life and death, growing and birthing, I opened the door to care. It's easy to remain detached from the *idea* of a dog; much harder to withhold care from the dog who sleeps at the foot of your bed, who waits beside you as you sketch the last of the morning light, who stumbles with pain as you wait for the damn vet to arrive. Who, in the final half-hour of her life, finds a way to rest quietly, head erect, on her bed in our sunlit living room, gracing Marc, Maggie and me with the time to say goodbye.

Sitting here in the cold sun, tears in my eyes, warm tea in my belly, pen in my hand, Embleton Mountain in view, I get it. A carrying capacity has limits; our capacity to care does not. Care fertilizes itself – even when it hurts. In my life, learning to care for one dog led to two dogs, and then to a child.

If we understand that the tools we carry – and the practices they engender – shape us, then each time we pick up a rock, a saw, a hand lens, a smart phone, we are making a choice about who we want to be, about how we will, in turn, shape the world. What we carry matters. If stone tools prompted our emotional connection to animals, what, I wonder, could pen and pencil cultivate with plants – those green beings who root the world?

I am a woman who misses her dog. A botanist who draws. An ecologist who gains comfort from the rhythm of a good sentence.

Today, no matter the risk, I draw care.

17. A Kind of Courage

LOVE REQUIRES VULNERABILITY.

You know this, even if you ignore it. That is, until you are jolted out of your complacency. Say, the first night your daughter, grown into the body of a woman but with the forgetfulness of a child, stays out until dawn, and you spend the night awake, rigid with rage, brittle with fear. But isn't this vulnerability the very thing that cracks you open? That allows you to know, in that first opening squeak of the door, what it means to be spared. That reminds you, once more, that your daughter's astonishing presence – fierce-spirited and loyal to a fault, yet apparently incapable of sending a text – started with two microscopic cells you never even saw, finding one another deep within your body. That, before you even knew her to be, the cells of her body had their own intent, dividing first this way and then that, following the directions of a coiled blueprint that, for all its stability, risked failure with each new division.

EARLY ON A SATURDAY MORNING, in middle June, I am standing roadside, en route to Botany Pond, wondering what counts as courage.

Around me, green hills gather violet shadows. High clouds linger in a blue sky still moist with last night's rain. Moments ago, as I closed my truck door, ravens called up a ruckus (*juveniles on a carcass?*), and two white-tailed deer regarded me steadily before ponging away, tails flashing. Now sparrows trill from a birch tree, and rough fescue rolls away toward the horizon.

All the world but me brims with potential.

Frankly, I'm exhausted. It's been a hard month, a hard year, a hard decade. Seventy years ago, ecologist Aldo Leopold wrote, "One penalty of an ecological education is that one lives alone in a world of wounds."[1] Today, I can only wish for Leopold's isolation. Spreading from philosopher to poet, from dancer to artist, from textbook to novel, ecological distress has become so prevalent it's earned its own name. Modelled after nostalgia, *solastalgia* refers to the pain caused when the places we love are lost to the wounds they carry[2]: when the fog that defines coastal forests disappears; when the familiar colours of summer are permanently shrouded in copper-covered smoke; when the four highways that stitch together interior and coastal BC simultaneously buckle under the force of catastrophic floods.

I fear losing place, but I fear even more *the time* it takes for us to notice its loss. Many of us, it seems, are surprisingly inept at perceiving environmental degradation. We gloss over the hurt and have trouble imagining those we never got to meet. Whether it arises from our faulty memories, or stories untold, it seems we are predisposed to assume that what we *experience* – rich or poor, intact or degraded – represents what always was. Ecologists refer to this tendency as the shifting baseline syndrome.[3] Today, botany, like most field sciences, carries more anxiety, more need to remember, than ever before. Is it fair, I wonder, to introduce students to botany's intricacies, when so many of its species hang in the balance? Is it fair to expect these students to remember what so many forget so easily?

Standing on the edge of the road, I need something to pull me away from my doubt. Something concrete and tangible; something outside my head.

An event map, I decide. I first learned to make these maps with artist Hannah Hinchman during those years I spent working as an itinerant botanist. These maps worry less about topography and more about what Hannah calls *events*. Anything that attracts the map-maker's attention – a flower thrusting its pigments skyward, a beetle ransacking another – counts as an event. Along a wandering line on the page, one event can lead to another, and then another, culminating in an explosion of observations made from one spot Hannah calls "a rapture."[4]

I know I could use a rapture or two.

Shouldering my backpack, I pull out my field journal, step away from the truck and fall into the roar of this grassland.

The crescendo is everywhere.

It's in the wands of grass shifting in the soft wind, the call of loons flying overhead, necks outstretched. It's in the orange rust coating the rose, the rounded mushrooms emerging from the dried discs of cow manure. It's in the warbles and elongate syllables of birdsong, the armada of mosquitos marshalling around my hands and face. Most of all, it's in the carpet of flowers flooding across the hills.

There's wild strawberry and chocolate lilies, penstemons and yarrow, Thompson's paintbrush and rough fescue, death camas and three separate vetches: American, timber and hillside milk. Each flower, I realize, has its defining gestures. The arc of common wild onion, sinewing upward like a snake tasting the air; the self-shading of rare bulbil onion, hiding its flowers in layers of sap green and dusky violet. The clambering of American vetch, with curled tendrils overtop its neighbours. The cheerful beam of orange arnica. The purple and red harlotry of larkspur and geranium.

Of course, there's more. Layer upon the plant life, the large bumblebee, whose name I don't know, trolling for pollen – the first one loyal to American vetch, the second visiting only larkspur. Record the butterflies: ochre ringlets for sure, a swallowtail, a little blue and the soft yellow flash of a sulphur. Don't miss the sap-green, hexagonal insect clinging to a faded balsamroot (*eating seeds?*), and then the birds — warblers and sparrows, cedar waxwings and swallows, swooping low over the meadow.

It's too much. Even for the simplicity of an event map. In this meadow, in early June, botany is not a series of events that can be noted, one after another, like beads on a string. This crescendo splashes into whitewater, overfills its bank. Even as I am caught in its torrent, I wonder if I will get more than ten steps away from the road before it's time to go home. I resort to listing, anything to capture one moment before I wash into the next.

It's not unexpected, this flood. Crowded into a narrow window of abundant precipitation and light, this surge of colour is something I ache for during winter's simplified palette. I also know how quickly it will fade. In June, I expect this grassland to wash with green, to dab with flowering yellow and pink and violet. But this – this is different. In more than a decade walking this grassland, I've never seen floral pigment *flood* across entire hillsides.

As I force myself to move toward Botany Pond, I make guesses about the events underwriting today's extravagance. Certainly, the last two years of relatively abundant rain helped build the reserves necessary to support today's abundance. Then a surprisingly cool spring that extended each species' flowering period. Further back, the shifting mix of seasonal precipitation and fire that favoured grass and flower, not conifer, at this elevation. And still further back, seeds blown north from populations persisting south of the glacial ice. The ice, itself once surging southward, creating the topography over which I walk: till and outwash, esker and drumlin.

But it's impossible to see today's exuberance without considering other, darker questions. Is this year's apparent exuberance just one more lesson in my understanding of this grassland's ecological variation? Or does my expectation of floral *dabs* rather than *floods* represent my baseline acceptance of an already impoverished world? Is today's extraordinary bloom the final echo of the world that was? Given the lag between atmospheric carbon concentration and Earth's climate, will the events underwriting today's bloom ever come again? For decades, climate scientists have been telling us that we will lose places before we see the evidence before our eyes. That we will walk sand beaches already lost to tomorrow's sea level rise, inhabit forests destined to be lost in catastrophic wildfire, live in homes and cities unprepared for next summer's heat. That Earth will still have to equilibrate, even if all of us stopped emitting carbon pollution today.[5] That today's increased temperatures are already feeding back into the complicated webs of carbon storage and decomposition. That, as ice caps melt, Earth is reflecting less, and absorbing more, solar radiation. That yesterday no longer predicts tomorrow.

I AM WANDERING a thin sliver of grass that stretches between aspen and conifer, when, abruptly, a sharp rattling call erupts, pulling me up short.

Just in front of me, two sandhill cranes lift from the ground. Adrenalin, and then concern, flood through my body. I don't want these birds to abandon eggs, or worse yet, chicks. Not when I know coyote and bear have their own young to feed. But the birds fly only 15 metres behind me before settling to the ground. One hunches low, while the other (*the male?*) stretches to his full metre-tall height and takes first one step and then another toward me. The other bird follows.

What, I wonder, should I do?

I don't have a bloody clue. I've practised responding to a charging bear, depending upon whether it's black or grizzly. I know what to do if a lightning storm breaks overhead. But *this*? No, in all my worries about the future, I've never anticipated being stalked by two birds, each nearly a metre tall, each with a beak as long as my pocketknife.

It'd be funny if it weren't so bizarre. A newspaper headline runs through my head: "Lone Botanist Runs *Afowl* of Avenging Crane." They won't forget that.

June's Roar

In the grassland, the roar of reproduction is everywhere. It's in the squadron of mosquitoes marshalling around exposed skin, the rounded shapes of fungal parasols, the abrupt appearance of sandhill cranes and the flood of floral pigments. This grassland tangle of limb and tissue is the only reservoir of the land's memory.

Each year, it is the roar rising again that safeguards and transforms the biochemical knowledge of what works. Preserved in the gene pool of cranes and coyotes and geraniums and butterflies and you and me, this cache of DNA is priceless.

Down in the valley water floods through channels of sand and silt and gravel. Up here, life floods through living tissue. This is the roar. Don't hesitate.

Studio Illustration: June's Roar

The cranes continue to advance, and I need to move. The birds are below and behind me. The terrain above me is steep and rocky, and not anywhere I want to go. I'm not comfortable turning my back on them, but I have no choice. I walk east, trying not to run. When I look behind me, the first crane seems to be breaking into the crane version of a gallop, wings slightly extended. *Jesus, what now? Maybe I'm walking toward their nest?*

In a swoop of too-large wings, one bird is up and over me, landing on the very ridge I'm headed toward. Short calls erupt from its location. *Is it calling a chick?* A quick look behind me confirms there is enough distance between the second bird and me for a downhill escape. I drop quickly, looking back over my shoulder often, losing sight of the birds once I've dropped below the crest of the next hill. With a good sightline behind me, I sit, not wanting to lose the sense of this encounter.

Majestic, stately, primitive, agitated, vulnerable, determined, fierce: adjectives fill my brain. None of them right; some of them contradictory – all part of today.

All pointing to one irrefutable fact: Giving up isn't an option. No matter the grief, no matter the fatigue. Not for me, not for any part of this green tangle of leaf and flower, muscle and limb. Life is hard-wired to recreate – to grow and thrust toward the light.

For a rare few like the pussytoes that grow nearby, reproduction is solitary, as cells inside their nondescript flowers use a form of virgin birth to set seed. For most of us with large bodies and long lifespans – chocolate lily, crane, or human – reproduction requires the extravagant cost of courtship rituals, the risky business of sex, the burden of caring for young. God help the botanist who gets between a crane and its contribution to the next generation.

In the Anthropocene, the world changes at breathtaking rates. But we are not the first to experience the grief of rapid change. We live, after all, on a planet where life constantly alters the conditions under which it lives. More than two billion years ago, a group of photosynthetic bacteria, cyanobacteria, started releasing oxygen – toxic to many of the species alive at the time, yet essential for the eventual evolution of larger cells, bigger bodies. Nearly 500 million years ago, when plants arrived on land, they began to release the organic acids from their roots and decaying bodies that forever rejiggered the cycle of carbon through continental rocks – causing atmospheric carbon to drop by an astonishing 90 per cent.[6] Through it all, life has suffered meteor impacts, colliding continents, and advancing, then retreating, ice sheets.

It doesn't matter; pulled by instinct, driven by hormones, life has no choice but to risk its own reproduction. There is no ark large enough, no vault deep enough, that can protect the extraordinary cache of DNA found in the living bodies of bacteria and fungi and delphinium and cranes and bear. In Botany Pond, in June, what matters most is life's capacity to rise up. *Natural selection* – the differential thrusting and growing and dying of individuals – is the only mechanism that can sculpt what works from what doesn't into species adapted to place. Yet nothing is guaranteed. Not when sex randomly shuffles the genetic code. Not when natural selection favours some over others. *This* is what's too big to fit on an event map.

That beneath the mad listing of today's species, there are countless failures: lives not born, lives ended before they could find a mate, set seed, raise young.

Biologist Stephen Jay Gould famously argued that if we rewound life's evolution, its replay would be completely different.[7] We will never know how natural selection, replayed, would interact with historical happenstance. We can, however, make predictions about how the Anthropocene will impact life's trajectory.

Just like climate, evolution lags behind rapid change. We inherit our bodies, our genes, from those who reproduced best in the previous generation. More than a century ago, Charles Darwin used farmers' well-known meddling in the sex lives of their crops to help explain how new horticultural varieties could arise. How by controlling who bred with whom, or who bred at all – what Darwin called *artificial selection* – plant breeders had sculpted weedy mustard (*Brassica olearacea*) into cabbage and broccoli, cauliflower and Brussels sprouts, kale and kohlrabi.

The shock of Darwin's day was worrying how these same processes, even without human interference – what Darwin called *natural selection* – could result in *new* species. That the stark differences of natural selection's variation and reproduction, life and death, could fill the world with chocolate lilies, death camas and balsamroot, warblers and sparrows and cranes. Today, in comparison, we worry far less about the origin of new species and far more about existing species' capacity to keep pace with our global meddling. In the Anthropocene, estimating the number of plant species facing extinction has "become almost an obsession."[8] Some compare the extinction rate in the Anthropocene to the time before (500 – 1,000 times higher!); others try to assess how many species are threatened with extinction (20 40 per cent!).[9]

Yet, even as botanists worry, we understand the numbers are flawed, riddled with ignorance and extinction's debt. Many of the world's most threatened plants are rare species, with small populations and isolated geographies. Many of these species, we know, will be lost before we can learn their names. Other species, especially trees that can live for centuries, may be functionally extinct – lacking the habitat or pollinators necessary for reproduction – long before they disappear. Their loss is a debt the world has yet to pay.[10]

Flawed numbers or not, no one doubts the cause of today's increased extinction rate. Habitat destruction, invasive species, climate change: all can diminish a species' capacity to successfully reproduce. All cause natural selection to lag further behind. Looking forward, we may not know the specifics, but on this everyone is agreed. The cost of the Anthropocene will not be borne equally among all species. Over the last 500 million years, the world's flora – estimated at nearly 400,000 species – has been sculpted through countless interactions with neighbours big and small. For nearly six centuries, botanists have gloried – whether by naming or collecting or painting or hypothesizing – in this extravagance.

Now the global meddling of the Anthropocene is most likely to favour the survival of the short-lived, the easy-to-disperse, the most human-tolerant.[11] Others, many others, will perish. How long can botanists remember? As the species we love – for their sensuous colour, their green possibility, their astonishing rootedness – stagger toward extinction, will we have the strength to stay by their side, to witness which of the world's wounds diminishes their bright light? After they are gone, will we know how to write the stories and draw the maps that will enable those who follow – animal and plant, microbe and rock – to track a different path than the one that's brought us to today? These are the questions that now haunt me, no matter the botanical abundance of any one season.

Yet sitting out here, in the moments before it's time to go, I return to my fatigue this morning, my nearly absurd encounter with the cranes. The awkward, yet indomitable, rising of these birds in open grassland of green and blue and purple and orange. Crane, daughter, flower, me. The risks we face take my breath away, but don't we only avoid risk when we avoid participating? And wouldn't therein lie the true death of the world?

LOVE TAKES COURAGE.

You know this, even as you fear it. Each coming year, as the world spins and tumbles in the unpredictability of the Anthropocene, you know its wounds will be bigger, deeper, scarier than you can anticipate. You know it will be those who are larger, who take longer to grow old, who have forsaken mobility's escape, who will be most likely to go extinct; who the world will be the most likely to forget. You know you will tire. You know you will worry. You know you will grieve.

And yet.

You also know there will be moments.

Moments when the world races full tilt toward the solstice, toward that light-filled maximum. Moments when floral pigment floods, sandhill cranes stalk, and butterflies flit through benevolence. Moments when there is enough sunlight, enough moisture, to allow the crescendo of a green world to roar with need and desire, glory and pain. Moments when you can't help but weep at the gifts you've received: legs, eyes, heart; the chance to be a mother, a partner, a teacher; the time to learn the names of plants, the space to draw their flowers.

Moments when the world rises up, insistent and urgent, demanding nothing less than its own recreation, and you understand, once more, what it means to be spared. Moments when you have no choice but to follow.

To rise up, to crack open – worry and grief, fatigue and loss, be damned.

18. Forest Refuge(e)

KLAMATH MOUNTAINS, CALIFORNIA

JULY 2019

DEAR AMERICA,

The forest from which I write is not one I know. It's dusty and dry, and supports the same number of conifer species – 30! – as my entire country. But here, along the banks of Mill Creek, in the rugged confusion of the Klamath Mountains in northwestern California, it's cool and shaded. Midstream, Maggie perches on a boulder, her nose in a book, her sandalled feet cooling in tannin-rich water, while Marc wanders away, up a narrow two-track. I sit on the moss-covered bridge spanning the creek and open my field journal.

Along the stream's edge, I see cousins to the plants I know 1200 kilometres to the north: *Aralia californica*, *Rubus vitifolius*, *Galium californicum*. In each name, a recognized genus flirts with a novel epithet. Together, the names chart alternatives to my flora, divergences born of differing histories of climate and community. This is not my forest, but it *is* the forest of Marc's childhood. Here, my husband is the botanical authority – drawing on an expertise cultivated during family holidays and college backpacking trips.

Even so, I'm a little startled to be sitting here with my field journal. My surprise does not reflect unwillingness. I may not know *this* forest, but 40 years ago, when my mother's spontaneous marriage dislocated my family south from BC, I learned about your generosity first-hand. I attended public school in Lincoln County, Montana, for nearly six years as an illegal immigrant, a child without papers. Even now, I wonder at how few questions were asked. Later, with a brand new resident alien card in my back pocket, I was supported – in university and then in graduate school – by your Pell grants and Stafford loans. Today, all but one of my diplomas bears the seal of your institutions.

America, I am not one of yours, but I married one you claim and gave birth to another. During the 27 years I lived within your borders, I watered your gardens, counted your plants and taught your children. You, in turn, gave me everything but a passport and the right to vote. And I declined to ask for more. Begrudging the decision that first brought me to you, I took your forests for granted until it was time to leave.

No, my surprise to be here, beneath this mixed canopy of incense cedar, arbutus, and bigleaf maple, has less to do with the man I heard a poet last week describe as the "orange-crested trumpfish" and more to do with my allegiance to the ecosystems found two state borders and one international boundary to the north. Fifteen years ago, when I returned to Canada, I thought I was returning home. But, home, I soon realized, was less *somewhere you went* than *something you practised*. Today, I know I practise home best when I attend to the botany of the Thompson Valley, with line and image, pen and paint, in my field journals. In the history of forests, 15 years is a blink of an eye, but in my journals, volumes 9–41 represent the most sustained attention I've ever given anything – other than Maggie, and maybe Marc.

Here's the thing. Today, the forests of interior BC, like so many, are shadowed by uncertainty. Count the losses. Eighteen million hectares of pine turned first red, and then dead, by the ravaging appetites of pine beetle populations escaping winter mortality in too warm winters. Millions more cut in the salvage operations that stubbled BC's interior plateaus with clear-cut. Add in the 2.5 million hectares of forest that perished during the 2017 and the 2018 wildfire seasons, the worst two on record. The vocabulary of climate change is convoluted – maladaptation, climate envelopes, productivity declines, nonlocal climate analogs, novel communities – but its implications are clear. The trees of my forest no longer carry the seeds of tomorrow.

Seen through the lens of time, no forest is stable or static. Fifty-two million years ago, California redwoods – a species whose range now stops near the California–Oregon border – shaded lakes near my home. During the last two million years, the forests of North America stuttered north and then south, with each ebb and flow of ice. The forest I'm sitting in might not be *my* forest, but at the height of the last glacial maximum, it – along with Mexico's Sierra Madre – was crowded with refugees from my forest.[2]

If BC's trees are maladapted to today's climate, then migration is again necessary. But no forest travels quickly. Fossilized pollen records suggest that the last time North American forests marched north, they travelled at a rate of 100–200 metres per year, 10–20 kilometres a century.[3] This time, it might be slower. Any migrant will face competition not just from other plants but from us. Will there be enough soil among our highways, cul-de-sacs and plowed fields for windblown and bird-carried seeds to germinate and survive? Or should we assist?

Not an easy question to answer. Assisted migration – first proposed in the mid-'80s as a strategy to save species endangered by changing climates – has exploded into contention. Some ecologists describe it as "ecological roulette," a strategy with no good arguments.[4] Others, cognizant of the risks of introducing species outside their current range but worried about the time it takes for trees to migrate, chart a middle ground, advocating for "assisted gene flow," where southern seeds are shifted north within a species' existing range.[5] Some worry less about managed introductions by scientists and more about "maverick, unsupervised translocation efforts."[6]

Maverick or not, assisted migration is happening. On Cortes Island, off the BC coast, artist and writer Oliver Kellhammer is field-testing California redwoods in woodlots and clear-cuts near his home.[7] Even the provincial policies governing tree planting on BC's public land – rules not known for radical thought – now allow western larch to be planted outside its historic range.[8]

I can't decide what to think. What reeks more of arrogance: assuming we know enough to shift an entire species' range, or leaving trees to struggle on their own, even when we know their misfortune has arisen from our maltreatment? Throughout human history, trees have fed us, supplied timber for our houses, warmed our hearths and provided our medicines. In return, we've planted some and cut many more – their fortunes largely ruled by how we profit. Globally, we've not been good for trees. Since the rise of our civilization, their numbers have fallen by nearly 45 per cent.[9]

Suddenly, it feels wrong to be taking another gift – even just shade – from a forest. I stand up and walk into the glare. Climate change's one constant is uncertainty. But maybe this is also its blessing. After all, moderating the unpredictability of tomorrow demands that we repair our lopsided relationship with the world today.

Vol. 42: Along Mill Creek

Last year, in 2018, a special report by the Intergovernmental Panel on Climate Change identified reforestation as one of the best options we have to limit global warming.[10] This year, a new paper ran the numbers. If we reforest nearly a billion hectares (there's still just barely enough land available), the growth of these trees could absorb and store more than 60 per cent of the carbon that humans have released into the atmosphere.[11]

The questions, of course, are what seedlings? To plant where?

I do not underestimate the complexity of these questions. Planting any tree is an exercise in hope, but planting trees, alone, will not mitigate climate change. We need to foster the web of interactions that underwrites *forest* – to plant seedlings where they can gossip with one another via extended fungal networks; where they can transform sunlight into sugar, and then wood; where they can expand, meristem cell by meristem cell, for centuries. Most of all, we need to plant seedlings where they can grow into the climate of tomorrow.

From around the corner, Marc appears with both hands full. Reaching me, he shares the names of the trees he has collected: Oregon ash, canyon live oak, giant chinquapin, black oak and incense cedar. With their long history in a warmer climate, these trees are part of Marc's flora, not mine. But, before we leave this forest, I pack leaves and branches into Ziploc bags and store them in the cooler. I want to take them partway home.

For the next two days, we drive north. With each passing kilometre, I relax into the familiarity of a known flora, even as I understand how climate change mocks the lines I draw. Before we cross the international boundary, I store Marc's samples in a relative's freezer in Bellingham. I can't let them go. Not yet.

With interactions spanning both generations and species, forests transcend time and continent. In doing so, they supersede most boundaries. Migrant, sanctuary. Refugee, refuge. Forests supply both. In the months to come, I will return to Marc's branches and leaves, taken from a forest I may never learn. Drawing their contours, I will do my best to imprint on their southern allegiance. America, these are not the trees of my forest, but they could be the hope of my daughter's.

With gratitude,
Lyn

19. An Unquiet Botany

I'VE LONG BEEN WARY of my garden. Unruly, crowded and small, it is, like many in North America, rooted in contested soil. It is, like many in BC, a novel ecosystem, inhabited by exotic species selected more by the Columbian Exchange – that great reshuffling of the world's botany – than by soil or climate. For more than a decade, I have tilled its soil, knowing the way we humans privilege some plants over others (think potatoes, wheat and corn) helped usher in the Anthropocene. For 16 springs, I relished its small, green uprisings, even as I dismissed it as part of the world we humans have planted into homogeneity. My body relaxed beside its flowers and ate its tomatoes and peaches; my attention as a botanist rarely lingered. That is, until first the native bees and then a novel virus arrived.

The bees sent their emissary to my office door: a spry, 68-year-old woman with an artist's eye, a gardener's patience and a grandmother's indomitability. We needed, she said, to count the bees. Kamloops sprawls across a continental flood plain, clambers up hillsides of sage and pine, and squats within a hotspot of native bee diversity. But nobody was paying attention. At the very least, she argued, we needed to know if bees visited the gardens we've planted. To understand how many of the 400 bee species native to our valley visit the flowers we have as a city, gathered from across the world, that each summer we irrigate into an extended bloom that persists long after the surrounding grasslands quiet into their seasonal drought. She'd help raise the money, she said, and organize the citizen scientists – more gardening grandparents. But could I find a biology student to compile the data and write the report?

Yes, I replied, but only if I got to count too.

Counting bees. Never for longer than 20 minutes at a time, no more than once a month – June to September – but anytime we allow ourselves to be schooled by others' desire, we are changed. And bee desire, as flowers have long known, is potent. What I had once planted for colour and texture, I now replanted for pollen and nectar and nesting material. I stripped lawn and annuals; replaced day lilies with licorice mint, irises with beebalm; and squeezed milkweed, yarrow, catmint and phacelia in the borders.

Plant it, and they will come. I rejoiced the first monitoring session when I counted more native bees than European honeybees. By my third summer of counting, I knew to expect the spring emergence of bumblebee queens, the June cloud of tiny, black *Lasioglossum*, and the episodic appearance of leafcutter bees throughout July and August.

In gardens, care composts care. That first summer of counting, the southern half of BC exploded into flame. In the limited time I could spend outside, I haunted my basil bed. *Ocimum*: a herb sculpted by Mediterranean soil and climate half a world away; *Megachile perihirta*: a leafcutter bee native to western North America. Two evolutionary strangers cultivating reciprocity with nectar and wing, their newfound devotion providing comfort even as smoke shrouded the predictable contours of my world. Any ecological rewilding – even that of garden microfauna – alters botany. Where once I prioritized plants primarily by their origin – native or exotic – now I wanted to know most about species' capacity for relationship.

Yet, even as I allowed myself to be schooled in bee desire, I neglected my garden's totality. When it was time for Aneka, the first student involved in our project, to list all the flowers growing in my garden, I wasn't even sure I could name them all. I spent the weekend before her visit immersed in technical floras. I felt the shift in relationship that occurs when any plant – even the rapacious weed with harebell flowers I've dug out of every corner of my garden – becomes known by name, *Campanula rapunculoides*. Several hours into our survey, with the back patio and three more beds still to count, I filled with a collector's pride when Aneka expressed amazement at my garden's diversity, but I never thought to compile my own list.

And then a new virus arrived. In March 2020, when we woke to a province-wide stay-at-home order, my unquiet garden became my *only* botany. With nowhere else to look, I took the list the students had compiled and added in my garden's nonflowering plants – the juniper and ginkgo tree, the ostrich ferns and the mosses that grow in roof shade. The result? More than 120 different varieties of plants. I was both surprised and reassured. Surprised, because this number far

1 Garden Shed

2. Back Beds

3. Lettuce Trug

4. Garage

5. Squash Bed

6 Back Patio

7. House

8. Raspberry Bed

9. West Side Beds

10. Herb Beds

11 Front Shade Beds

12. Vegetable Beds

13 Front Flower Beds

14. Verge Beds

1104 PINE ST

Illustrated Map: 1104 Pine Street

exceeds the botanical richness I'd tallied in similar-sized grassland plots in Botany Pond. Reassured, because the number felt abundant, more than enough to keep me occupied, even if quarantine lasted the entire summer.

During the first several ambiguous weeks of quarantine, there was no denying my garden's comfort. In March, snow still threatened, but by the second week of quarantine, the snowdrops and crocuses, and the new dwarf daffodils were already in bloom. Confined in place, I remembered to fill the bird feeders and was, for the first time ever, home to watch the songbirds return as the first queen *Bombus huntii* crawled woozily from one crab apple flower to another. Throughout April and May, as we worried about the loss of jobs, and hospitals filling and those we loved dying alone, the robins and varied thrushes, the flickers and song sparrows, the yellow-rumped and Townsend's warblers, the house finches and white-crowned sparrows, the hummingbirds and evening grosbeaks returned. They came to kick through last year's leaves, to drink from the red tube I filled with sugar water, or to eat the peanuts and millet and sunflower seeds in the feeder. One morning, I woke to the sounds of a mallard pair, lured from river into garden by spilled seed. In quarantine, I watched the emergence of at least four different bumblebee species, two leafcutter bees, one bright shiny green *Agapostomen*, and more sweat bees than I could distinguish.

Here's what I never anticipated: that with each new arrival, each successful emergence, comfort would jolt toward loneliness. But it did. It's not that I hadn't worried before. But in the South Thompson valley, the uncanny risks of the Anthropocene – frequent fire, extreme flood and heat domes – tend to threaten whole ecosystems, not just individual species. My worry before quarantine had always been in the company of sage and pine, junco and coyote. Isolated in my garden, I couldn't escape the comparison between the vulnerability of my species and the exuberance of others. Of course, it wasn't just in my garden. Reports filtered in from across the world: wild turkeys strutting through downtown Montreal; Himalayan mountains newly seen in skies cleared of pollution; bird and whale singing free of engine drone.

The implications were clear. The Anthropocene always has, and always will, privilege some at the expense of others. In retrospect, the real lesson of our *anthropause*[1] was the opportunity it created to link vulnerability with indifference. My loneliness, after all, was but a tiny sliver of the indifference faced by others. For too long, the Anthropocene's story of separation – nature versus culture, mobile versus emplaced, native versus exotic – has given me an excuse to look away. I accepted my garden's comfort without ever asking who it favoured. I worried about other people's plant blindness, even as I failed to comprehend the complicated lives of those who grew in my own garden. To consider the climate cost of the supply chain that brought their bodies to my local garden store, to imagine the hands that had gathered their seeds, the distance their genomes had travelled. During lockdown, it became clear: How could I re-story my relationship with the plants of place, to learn from their possibilities, while ignoring those who knew me best?

In July, I harvested garlic I'd planted months before I knew how to pronounce "coronavirus," and watched bumblebees ferry pollen from flower to nest. In September, I used rafts of brown-eyed Susan and *Echinacea* as a backdrop for my remote teaching. In October, I saved the last tomatoes from a freak early snowfall. In December, as rain fell, COVID-19 numbers surged exponentially, and I raked leaves from still-too-green garden beds, wondering if grief was my garden's new crop.

In retrospect, I was right to be wary. Nothing in an Anthropocene garden is simple.

But what if the path to resilience depends upon our willingness to move past indifference into relationship? To accept both its worry and joy, desire and vulnerability? No plant, after all, roots tangle-free. Below ground, the *radicle* hope of an embryonic root survives best in complication. Carbon-rich molecules squeeze from root cells and, in doing so, solicit the unknown multitudes of the soil microbiome. In exchange, plant growth is modified, defended, orchestrated. Above ground, relationships are no less intimate. *Gallardia* pollen packs between sweat bee bristles; beebalm plays host to powdery mildew; *Rudbeckia* petals dimple beneath the predatory weight of ambush bugs. My body, too, is implicated. Fingers prune and weed; thigh digs and mouth tastes. Small to big, microbe to plant to animal: it's not numbers but reciprocity that will restore the world.

Desire, vulnerability. Bees taught me the first, a virus the second. Both refuse indifference.

I wonder if learning to live in place begins only when we accept both the potency and vulnerability of *all* our relationships. In both biology and culture, interactions – positive or negative – drive change. Bodies transform, lineages diversify, attitudes alter. And, within any ecosystem, change ripples. By the time I germinated the seeds for my second season of pandemic gardening, I did not doubt the ripples of inequity that had spread from the SARS-COV-2 virus – that microscopic ball of RNA and spiky proteins that still haunted our species. But I wondered how long it would take my community, as a collective whole, to make the link between our own vulnerability and the precarity of others? To identify human indifference as the driving force that thinned the world's trees by half, that transformed our green neighbours from kin to commodity, that allowed our numbers to first rise exponentially, and then travel endlessly, bypassing the ecological barriers of time and distance that once constrained the risk of new pathogens, of unwanted relationships.

Caring or complicit? When it is time to start the timer on my next bee survey, to plot how to attend to the riot of flowers in bloom – *Echinacea* and *Phacelia*, licorice mint and *Ocimum* – I know my garden is both. Its flora, like most, is still too similar to gardens across the continent, still too indifferent to original inhabitants of this valley: those specialist bees and native wildflowers that, despite my coaxing, have been so far unwilling to enter its beds. Yet, in the next 20 minutes, each bee I tally will help erase indifference. Each plant I encounter – many newly arrived, a few long in place – will help root me in a web of indebtedness. Each survey minute that ticks past will be one moment of linearity in my garden's circularity, an understanding of time that cycles from one season to the next, from seed to leaf to compost, and then back again. In the next 20 minutes, each new act of intimacy – human thumb to leaf, bee abdomen to pollen, fungal thread to epidermis – will envelop me in a vegetable crucible of change.

My garden *is* complicated, but how many ways can it cultivate care?

20. Straddling Worlds

START WITH THIS: This is the garden that lichens built. Found across the road from my university's field station in the Upper Clearwater Valley, two hours north of Kamloops, this garden is not mine, although I've eaten its peas and walked its paths often. Neither have I tilled its soil, though lately I've begun carrying its carrots home.

Lichens first brought me to Edgewood Blue when I was a graduate student. I spent a long weekend sleeping down the road in my truck in the provincial campground-now-turned-parking at Spahats Falls, driving up by the day to peer through microscopes at the startling morphology of *Cladonia* species. Since moving to Kamloops, visits to the provincial park, Wells Gray, that cradles this valley on three sides have been my most regular migrations – at least one weekend every fall, if not more, and at least two weeks every other spring to teach an ecology field school. My Wells Gray field journals, unlike those I carry everywhere else, are numbered separately: Volumes 1–8. In my last art exhibit, the lessons I'd learned from this park's poor fens, wilderness lakes and old forest were second only to those I'd learned from the grassland matrix of Botany Pond.

But now, on a middle May morning, I'm in Wells Gray for neither lichens nor park. Earlier this month, I'd planned to be here co-teaching a workshop for early-career naturalists from across BC. But, without a vaccine, COVID-19 cancelled the workshop – like it did my field ecology course last spring. In the wake of yet one more cancellation, I ached for more botany than my small garden could provide. But, with the world so disrupted, it felt complacent, maybe even inappropriate, to fall back into the uncomplicated comfort of species long in place. Instead, what pulled at me was the garden across the road from my university's field station.

One I've walked through for so many years without much notice. The garden of Edgewood Blue.

First crafted from the interface of wetland and forest more than 40 years ago, the stony foundation of this garden rests atop Pleistocene flood basalts once eroded by glaciers, and then layered with fine silt that pooled in the wake of melting ice. For many years, this garden was but a few beds outside a small house in the woods. But then, our field ecology class began, each morning, to hear the intermittent rumble of an excavator beneath the song of kinglet and warbler. For the next year or two, traversing this garden to the forest beyond meant navigating the excavator's chronic presence. Each time we saw this machine – its squat and yellow form unbalanced by its single, double-jointed shovel – it appeared as precarious as it was loud.

Yet, even as I worried about angles of repose, the excavator's roar imagined new possibilities. With each scoop, it cleared shrubby carr to expose the open water that underlay this wetland. With each shovelful, it inverted floating wetland bottom into water-crossing dike and garden bed. Today, this wetland, now called Sky Pond, is an open expanse of reflected blues and greens, punctuated by cattail and lily pad, floated by cinnamon teal and mallard, swum by beaver, otter and muskrat, flown by swallow, harrier and snipe. Collectively, the cultivated beds of Edgewood Blue's garden now encompass nearly 500 square metres of deep soil. Some beds are shouldered into hip-height by moss-covered basaltic boulders too heavy to lift by hand. Some sprawl across hilltop or island. Still others are dug into the edges of Sky Pond.

Right now, in the fine possibility of a May morning, as I wander its paths, it's clear this garden, like many, erupts on the margins of hard-edged structures built by and for humans: a house, a guest cabin, three woodsheds, a greenhouse, a gas shed, two outhouses and one Open Learning Centre (the door of which you enter only to find two sides completely open to Sky Pond and surrounding forest). Like those in my garden at home, the plants of this garden are a novel community, but this community supports a higher percentage of species rooted in place. Certainly, apart from the Melba apple next to the greenhouse, all its trees are native. Over the last five years, the trees most vulnerable to fire – Douglas fir, spruce, subalpine fir – have been cut in favour of the more fire-resistant canopy of aspen and paper birch. Beneath its trees, this garden, at last count, holds more than 1,000 different

nuthatch eeks, • yellow-rumped warblers walk over
hummingsbirds buzz • rabins walking.

gooseberry

Between the Ribes,
the Hymenopteran drama oucks
me on; two bumble bee species
and a wasp, not necessarily on
good terms.   But I know that
it was the bees, the wasp
that first alerted me to
the presence of Ribes
flowers. Without the drama
I would have missed
this floral abundance.

black currant

G A R D E N
edgewo
10:45 am – 7:45 pm

Curtis says that he felt cooped up
when he first came to Edgewood, until he
started gardening, fermenting, making birch
syrup. For all its familiarity as a cliche,
I think about how rooting occurs through
doing. It is surprising, Curtis says, that the
garden here roots him here as much if not more
than the wildlands.

Polytrichum
male & females
on the raised
beds.

frogs calling, rain falling on tin roof

Wells Gray Vol. 5: Garden Walking

itat, • ravens carooking • sandhill cranes in the distance
catcher calling • red-wing blackbird "doo·doo·doo·doo"

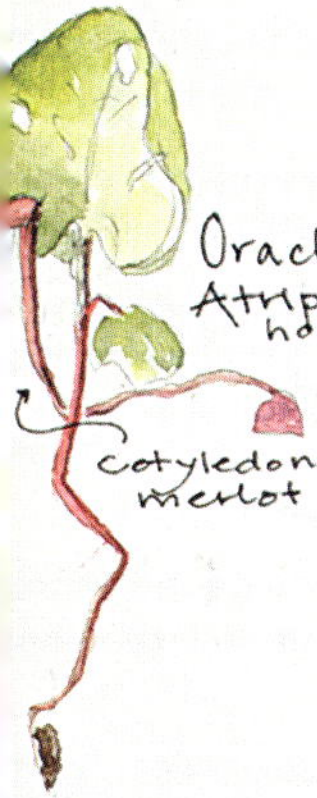

Orach
Atriplex
hortensis

cotyledons dense
merlot

Dark clouds advance south from up valley. I
ignore the raindrops for as long as I can before
I retreat to the Open Learning Centre.

Gardens embody our most intimate plant knowing.
I've long enjoyed the Ribes bed for its architectural
presence, mounded by rocks into hip height. But
for Curtis, the bed is difficult, hard to access
and all flat on top. Coming back from a walk,
Curtis estimates there is 500 sq. meters of garden beds
found adjacent to pond, greenhouse, house, even
some now out on Story Island.

W A L K I N G

Taraxacum   Pseudoroseum

o i n e   May 20, 2021
alternates with short, dark showers.

ling & scratching in the garden bed

plant varieties. Yet this garden, like most, says as much about its makers as it does about its plants.

This garden has grown most from the minds of two makers: Trevor Goward and Curtis Björk. Neither holds a PhD, nor occupies a faculty position, yet both have published more botany than I ever will. Both advise governments on the status of rare species; both serve as curators for one of western North America's largest lichen collection. Both have spent years learning to think not just *about* but *alongside* plants – their form and variation, their relationships with each other and their communities. For years, when ecology students have marvelled at the place Trevor and Curtis inhabit, my response has always been to tell them how Trevor's taxonomic expertise first gave him the "lichen dollars" to purchase the property he named Edgewood Blue.

"Choose something, anything, that you love enough to obsess about," I've told students, "and eventually people will pay you to do the work you love."

The length of Trevor and Curtis's tenure in this garden, like their age, differs by two decades, but both now begrudge any time spent away from it. Together, these two root deeply in the garden they share, but they also, I will learn over the course of several garden visits, disagree about the ultimate shape their garden should take.

CONSIDER THIS: lichens, like gardens, straddle worlds. In mid-June, I am awake early and about to embark on the two-hour drive that will take me from the sagebrush steppe of Kamloops to the interior rainforest of the Upper Clearwater Valley. As the only early riser in my house, I pop my phone's earbuds into my ears so as to not wake anyone while I listen to CBC Radio's weekend art show as I assemble the last of my gear.

Unexpectedly, I hear Trevor's voice in my ear. "Lichens," he says, "are like koans. The sound of one hand clapping."[1] He's speaking to the host – a woman I trust for her kind voice and careful questions, but who hasn't, I suspect, ever interviewed a lichenologist. In the slight pause that follows Trevor's words, I swear I can hear the debate in this woman's mind. Finally, she takes the plunge. Will Trevor explain?

All lichens, says Trevor, are individuals. They grow, have sex, reproduce, senesce, die. Trevor tells the radio host he's been sorting out the taxonomy of BC's peppered moon lichens   so named for the granular outgrowths of their upper surface and the crater-like impressions on their undersides. Each lichen is named, like our own species, following the rules of scientific nomenclature. But, unlike us, each one of these lichen species is birthed in relationship, not across the dichotomy of gender but across the yawning gulf of taxonomic kingdoms.

Freddy Fungus took a *lichen* to Alice Alga. Botany students use this mnemonic to help remember the rather extraordinary, always intimate, encounters that occur between the unrelated taxa contained within a lichen thallus. For many years, we thought each species of lichen included at least one alga, sometimes a cyanobacteria, and a *single* fungus. Several years ago, in a scientific publication that splashed across the world, one of Trevor and Curtis's closest lichen colleagues proved that a tree lichen, also known as edible horsehair, found in the forests not far from Edgewood, included at least two very different fungal species. One with tiny white threads; the other round and globular, a yeast.[2]

For an entire summer, Edgewood's garden buzzed with excitement.

Here's another full complication: Some, perhaps most, of the species of algae, cyanobacteria and fungi that create lichens can also grow independently. Yet none that do so look or act anything like a lichen. Every lichen, Trevor says, is an emergent property, something other than the sum of its parts. Lichens, Trevor says, are both individuals and tiny, self-sustaining ecosystems, built from the interaction of multiple lives.

I've heard this before. I have multiple pages in my field journals filled with versions of this explanation. I have sketches of fungal hyphae enveloping an algal cell – what Trevor calls the "moment of enlichenment." I've watched students hold the thin hairs of *Usnea* or *Bryoria*, the pelt-like thallus of *Peltigera*, the erect clubs of *Cladina*, in their hand and try to comprehend their complications. I've listened to the inside jokes told by a gangly group of lichenologists who came to sort through their species identifications with Trevor and Curtis. I've transcribed the words of poets and writers, naturalists and artists, linguists and philosophers, as they explore lichens for metaphor and meaning with Trevor. I know how thoroughly lichens have, in Trevor's words, colonized his mind. I know where this is going, and I'm racing ahead, anxious. What lichens have taught Trevor might be a bit hard for CBC's listening public to digest along with their Sunday morning waffles and coffee.

On the radio, Trevor says what is true for lichens is true for the rest of the world, for *Gaia*. No individual can exist in isolation; any lichen teaches us how the part must respect the whole, and the whole must respect the parts. The resilience of anything, including that of lichens and the world, emerges only through the synergies that arise between the parts and the whole. If any one part forgets the whole, the entire lichen – its fungal cortex and medulla, its algal solar cells, its rhizines and soreida – is risked. I know Trevor believes that our increasing lack of attention to the whole, our willingness to prioritize our needs over that of our neighbours, imperils us all.

Together, Trevor and Curtis have more than 50 years of residence in this valley. Talking with either man can feel like inhabiting a liturgy of grief. Little in this valley has been unaffected by our cultural inability to respect the whole. Some lichen species have disappeared entirely from lowland portions of the valley, while others now persist only in their core habitat – for now. The same is true for the mountain caribou of this valley: Today, these animals number barely more than 100 individuals. Extirpation is probable. Last year, two nearby creeks flooded catastrophically, even as glaciologists predict the demise of BC's glaciers.[3] Above all else, summers in this valley grow in precarity, wreathed in smoke, threatened with flame. Where once an excavator dug for increasing species diversity, now trees are thinned and buildings reclad in fire-resistant siding. Seen through lichen eyes, we humans have birthed a world uncanny with risk and dark with loss.

I know all this. But when my mind stops racing ahead and returns to listening, the radio host has already moved on.

FEEL THIS: every garden, every lichen, is negotiated space. It's late afternoon in June, and I am on the deck – once wood, now cement – overlooking Sky Pond and its adjacent beds. Below me, Trevor pulls *Equisetum* stems from the pond margin, all six foot five inches of him bent in concentration. Curtis has just left for the monthly meeting of the Upper Clearwater Valley volunteer fire brigade, his tall, thin frame clad in official uniform: bright red shirt, blue pants.

It's too easy, I think, to assume that all negotiations necessitate words. But from my vantage point above Edgewood's lower beds, I wonder if the language of negotiation can't also be more nimble, more fluid than that which can be expressed in words. In gardens, don't we argue as much with spade and trowel, inadvertent seeding and differential weeding (as Trevor calls it), as we do with verb and noun, supposition and proposition? How quickly, in my mind, the contentious discourse of *argue* tracks in behind *negotiate*. In gardens, can't love and sorrow, colour and fragrance, matter as much as disagreement?

Of course, some negotiations transform into consensus more easily than others. Earlier this afternoon, a silky black bear and her yearling cub ambled past, content in the moment to wander in full view, but on the other side of the tall, black, wrought-iron drift fence that usually settles any debate about which parts of Edgewood Blue are open access. Water-crossing dike, built purposely as wildlife corridor, *yes*; vegetable beds and fruit tree, *no*.

Here's where Trevor and Curtis still disagree. In the garden, Curtis grows for diversity over aesthetics, Trevor, for aesthetics over diversity. Curtis says that, when he first came to live at Edgewood, he found its plant hardiness zone, 3B, nothing less than depressing; it's location along a dead-end road, confining. But learning to garden at Edgewood has allowed him to travel less, root more. In the midst of writing a new flora of BC's flowering plants, Curtis now grows alpine *Poa* species on the shadiest, coolest island within Sky Pond. Rather than trying to sort out the morphological variation of these grasses based on snapshot encounters with them in the field, Curtis cultivates them so he can monitor their variation from first green leaf to full emergence, from seed set to senescence. In Curtis's garden, he is the keeper of lists, the arbiter of species boundaries, the protector of difference.

Trevor, in comparison, says he would be happy with only 80–100 species in the garden, if they formed masses of colour and texture. For Trevor, diversity matters most in how it serves as a gateway, a portal, into wildness. In Trevor's garden, weeds are both opponents and allies: plants to be pulled, and the wild making itself known. Along the margins of Sky Pond, *Equisetum* stems normally stretch tall, but when stressed with repeated pulling, their growth turns horizontal. Such shifts in growth pattern, Trevor says, are "what plants look like when they're thinking." Weeding reminds us that any ecosystem – the human body, a lichen, a wild valley – always learns from lived experience.

Compared to lichens, garden negotiation feels easy. No group of organisms challenges my understanding of the world more than lichens. More than 30 years ago, Trevor described lichens as fungi that had discovered agriculture. Since then, he has renounced this description – all too aware of how it privileges the perspective of the fungal partner over that of the algal partner. Trevor's long-time friend, poet Don McKay, calls lichens a "metaphoricity become incarnate, as though an abstraction turned up, plain and palpable as a dandelion in your backyard."[4] Metaphor and ferry, Don says, share the same root. Trevor has since taken to describing the body of the lichen, technically the *thallus*, as an infinitely detailed transcription of a very long conversation between the lichen partners. Much of conversation that is recorded revolves around carbon: how quickly the alga community can spin carbon dioxide into sugars; how rapidly the expanding fungal hyphae will drink sugars down; the abundance of sugars relative to nitrogen – that other most limiting nutrient in terrestrial growth. The tongue-twisting nouns lichenologists use to describe the different parts of the lichen thallus – pseudocyphellae, papillae, scabrosities, scleridia, tomentum, rhizines, holdfasts, pilema, stereomes – are, in Trevor's view, less nouns than verbs working to boost, distribute or store carbon within the lichen.[5]

Merriam-Webster says that to *negotiate* is to confer with others, so as to arrive at the settlement of some matter. Confer. Arrive. Settle. Don't all of us, every day, in each breath we take, in each minute extension of hyphae around photosynthetic cell, in each seed planted (or not), negotiate life from death?

WONDER THIS: Is learning to live in place – what some call *indigenization* – an emergent property?

Indigenous scholar and poet Jeannette Armstrong describes "indigeneity to a place" as an "adaptation over a long period to the realities of that place…in ways that promote interdependence."[6] From a different continent, philosopher Martin Lee Mueller writes, "Indigeneity…involves gleaning thought-structure from the patterned animation that is specific to each place."[7]

In middle June, the upper garden beds beside the greenhouse are punctuated with seed packages inverted onto stakes. It's already past 6:00 p.m., but the sun is still two hand widths above the green horizon. I tiptoe through the rows, cataloguing Curtis's seed selection. All the packages I can find are from the same seed company – Baker Creek Heirloom Seeds, from Missouri, of all places – and include a blue corn, two beet varieties and three different beans. Across the path, unnamed tomatoes and peppers, chives and strawberries, a row of snap peas clambering up chicken wire, and blooming blueberries gulp the last of the sun. A black-bottomed bumblebee visits at regular intervals, ferrying pollen and nectar from blueberry to nest.

Last month, I spent two days walking with Trevor through the forest that stretches from the margins of Sky Pond to the border of Wells Gray Provincial Park. Trevor calls this area *Island Earth*. It's not what most would understand as an island, but it is, like most places on our watery planet, bounded by water on all sides: one river, several creeks and a wetland complex. Island Earth isn't a large area – only four square kilometres – but the printed map Trevor handed me when I arrived was crowded with names: Whisperwood, Blue-Rib Canyon, Snowborn River, Wind Falls, Ravenfall, Cathedras. None of these names match any you'd find in the same area on the official topographic map, but with the recent floods, the easiest path into Island Earth now begins and ends in the garden of Edgewood Blue. During those days of walking, my field journal overflowed with notes from Trevor, all of them testifying to a long residency in place.

Yet the most startling lesson I learned came not on the trail but after the fact, through the juxtaposition of Trevor-in-the-forest with Trevor-in-the-garden. Within Island Earth, when Trevor wants to talk, he halts on the trail and turns to you, leaning his height on a thick walking staff. In the garden, Trevor begins nearly every conversation by first excusing himself to bend close to the ground to continue weeding or mulching. On the trail, Trevor, paraphrasing linguist and poet Robert Bringhurst (another friend of Edgewood Blue), says, "Wilderness is where you don't have to weed."[8] There's no doubt that the logic of gardens like Edgewood is less permanent, less resilient, than that of wilderness, yet I know few humans who live in wilderness. If we are lucky, we live in gardens where we can't escape the consequence of our weeding (or lack thereof). If we are even luckier, we live in gardens, like Edgewood, that open to the wild, sometimes even to wilderness.

There's no doubt in my mind that too many of us have thought or forgotten ourselves out of the *whole*, but there is no guarantee that thinking alone *will* or even *can* rebuild such relationships. Long ago, Aristotle wrote that anything we have to learn to do, we learn by the actual doing of it.[9] Doesn't rooting in place, as an emergent property between organism and environment, depend upon negotiation, upon the give and take of experiments? Leaving the upper garden beds, I think of Trevor's weeding, his evolving description of lichens, his relentless observations of both their form and distribution, his extraordinary capacity to hold their koanal existence in his mind. I think of Curtis's alpine *Poa* species out in Sky Pond, growing well below their native elevations, the heirloom seeds imbibing water filtered through the stone of this valley, the carrots Curtis sends me home with, his indefatigable list making. Is there any doubt that both of them, even when they don't agree, garden in support of the whole? That their bending of muscle and bone, mind and heart, year after year, has cultivated interdependence? That their willingness to imagine life outside their own skin, whether in fungal thallus or garden carrot, has allowed this place called Edgewood Blue to shape how they think and who they are as people?

I'M BACK ON THE DECK, wanting to see the slanted evening light across Sky Pond and the garden. The mosquitoes are gathering. My mind feels exploded with possibility, my right hand tired from drawing. What form, I wonder, could *all* of our gardens – big or small, near or far, ordered or rambunctious – take if we learned to see them less as objects and more as conversation? If we understood their species – native and newcomer alike – as *parts* in dialogue with the *whole* that supplies our water, houses our pollinators, and composts our debris? Tomorrow my ankles will be ringed with mosquito bites, the molecules of my red blood cells already incorporated into other lives. I know I will need to return to this garden again and again; that the complications of Edgewood Blue will colour yet more pages in my journal; that I will walk with Trevor further and further into the woods.

This is, after all, the garden that lichens built.

21. Cultivating Troubled Soil

LITTLE ABOUT THIS SMALL PATCH of green is what I expected. Butler Urban Farm is neither pretty nor pastoral, neither owned nor gated. Rather, it is a farm awash in a sea of warehouses and trailer courts, car dealerships and auto part stores. It is a farm whose crops root in ground only recently reclaimed from bindweed; whose food is only ever gifted, never sold. It is also a farm that sits within the same floodplain as the Kamloops Indian Residential School, where, three weeks ago, the bones of 215 children were found in unmarked graves. Where now 215 crosses, dressed in shirts of orange and red, stand mute along the highway, bearing a nation's grief.

Kneeling in a potato patch just across the river from those crosses, clad in a long-sleeved shirt, shaded by a floppy hat, I wonder if the most important crop this farm grows is not root or shoot, flower or leaf, but its willingness to cultivate troubled soil.

It's just past 8:00 a.m., late June 2021. For the past several days, the Environment Canada app on my phone has been emblazoned with a bright red heat warning. Tomorrow will be the first of six days with temperatures climbing above 40°C. Climatologists have been using words like record-breaking, unprecedented heat dome. Here, in Kamloops, summer temperatures bouncing close to 40°C are not unexpected. But never a whole week of plus 40°C. Never plus 40 in June. This is heat made rageful by my species' profligate use of carbon. I know this might be the last full morning I can take my field journal outside, and I'm sketching as fast as I can.

From my place in the potato patch, I hear the *clu-chunkk* of a train coupling, the whir of cars speeding along the street elevated five metres above me. Beside me, house sparrows raise a ruckus on a chain-link fence. Off to the left, a trim woman I've only just met, Sandra, gray-haired and recently retired from the

Kamloops Food Policy Council, bends over, weeding her rows of corn. She comes once a week, she said. Grateful for the opportunity to harvest beyond her small yard. From where I kneel, potatoes mound in straight rows, leading to a terrace of flowers, a peach tree, a Saskatoon bush.

As the sound of traffic builds, I wonder how many people speeding past notice our bent forms. I know this farm is easy to miss. I certainly did on my first visit. That day – less than a month ago – Maggie had left our house on her bike, pannier stocked with water bottles, floppy hat and gardening gloves. Then, near noon, she'd called. Weeding had taken longer than expected. She had to get back to our house for a video meeting sooner than she could ride.

Could I give her a lift?

As the crow flies, the farm is less than five kilometres from our house, but it's in a part of the city I don't know well. If the name Kamloops originated from the Secwepemcstín word Tk'emlúps, for "meeting of the rivers," the population of this city – numbering just over 90,000 inhabitants – now sprawls far away from the confluence of the North and South Thompson rivers. Unlike Kamloops, the name Thompson has shallow roots in this valley. It was affixed from afar; one stranger to this land honouring another. Two men sent to different parts of northwest North America to find passage for furs and trade goods and more Europeans. Geography and history books report this naming as a matter of fact. Simon Fraser named the Thompson River during his 1808 exploration of the Fraser River to honour David Thompson.[1]

I know cultural mores change with time; I know it's easy to criticize from 200 years distant. I can only imagine the courage it took to push off from the bank, canoes loaded, into a mapless descent of a big river. But still, today, I wonder about that moment of naming. It was in June, just as the global climate was recovering from the ecological consequence of European colonization of the Americas. From the disease and genocide that had emptied Indigenous fields of 90 per cent of their peoples, the growing tree canopy that then colonized these empty fields, gulping carbon dioxide, likely ushering in the chill of the Little Ice Age.[2] What did he think, this man born near Bennington, Vermont, buried near Cornwall, Ontario, paddling through open country of sage and rabbitbrush, when he encountered yet one more tributary coming in from the east? Was naming an obligation or a privilege? I know he paddled alongside 23 other men, and that the party was guided through Secwepemcúlecw by a Secwépemc Chief from Soda Creek,

Cllecwúsem, who undoubtedly knew the first names for not just the rivers, but for the hillslopes, the iconic rocks, the plants.[3] When Simon Fraser chose Thompson as a name for this tributary, did he wonder what stories, what jokes, might ride alongside the name of this river in languages other than his own?

Two hundred years later, this is what I know: The two men involved in the naming of the Thompson River never saw the extraordinary confluence that sits at the heart of its watershed; never saw the grayish, silt-laden waters of the North Thompson, undammed and unsettled, rushing in from the north to butt up against the crystal green waters of the South Thompson, lake-filtered and west-running. Never listened to fluent speakers of Secwepemcstín chuckle together as they tried to explain the pun embedded in Tk'emlúps.

Yes, said Ron and Marianne Ignace – gray-haired Secwépemc Elder, and German-born scholar of Secwépemc plants and language, standing together at the front in a lecture hall at my university – Tk'emlúps means "where the water meets." But, when seen from the plateau above, the curve of this confluence echoes the curves seen on certain human bodies when seen from behind. One word, two meanings. Meeting of the rivers; buttocks.[4] Ribald perhaps, but for those who live in place, why wouldn't land and body joke with one another?

David Thompson may never have learned the double entendre of Tk'emlúps, but few doubt the impact this man's career had on the history of my home. His great map of western Canada – measuring more than ten feet by six feet – is rich in cartographic detail. It was also a critical tool in the ongoing survey that helped imagine Canada into being,[5] in the transformation of the land into something you could own, rather than something to which you belonged. Maybe it's not unexpected that the rivers named after this man still carve lines of privilege and opportunity, reserve and municipality, across my city.

South of the confluence in Kamloops, neighbourhoods tend toward historic or gentrified. Many homes are small – bungalows built in the first half of the last century – but there are no trailer parks. Instead, a French immersion grade school and my university attract students from across the city and the world.

North of the confluence, neighbourhoods divide by more than just economics. East of the North Thompson lies the land recognized by the Government of Canada as belonging to the Tk'emlúps te Secwépemc – one portion of the tiny reserve granted to a Nation whose unceded traditional territory spans more than 145,000 square kilometres, an area equal to that of New York state.

Cultivating Troubled Soil
—

To the northwest of the confluence, where the rivers flood most frequently, lie the crowded streets of North Kamloops. When my family first arrived in Kamloops, our well-coiffed real estate agent immediately dismissed North Kamloops as "not suitable for us." That evening in the hotel, Marc read about the rise of the sex trade on the North Shore, the IV needles found in its vacant lots. In two days of furious house hunting, I don't remember crossing north of the confluence once.

A month ago, when Maggie texted me the address of Butler Farm, I knew it sat on the North Shore's finger of floodplain. But Google Maps had yet to pin the farm's location, and the street address took me to the corner of the farm opposite its driveway. In my truck, the Google woman had to tell me to turn around twice before I found the right arterial, the southside of which fell steeply down to a trailer court, and then into the green space where my daughter waited.

Turning down the dirt driveway, I was surprised to recognize the thin profile standing next to Maggie. It was Kevin Panewich – a recent graduate of my university, a political science student who sat in on the first few sessions of one of my courses, who came with the class across the North Thompson when we visited the Secwépemc Museum and Heritage Park, who was hugged by the Tk'emlúps education specialist who was to guide my class through the museum and garden. A young man whose perspective I missed when time constraints forced him to drop my class.

He was, he said, the Butler farm manager. I had to go – I sensed the moments ticking by – but before we left, Kevin invited me back. He'd be delighted, he said, to show me the farm.

THREE WEEKS LATER, field journal in hand, I am frustrated with the curve of a potato flower's exserted stamen when I hear Kevin's voice in the distance. Earlier this morning, Maggie had guided me back to the farm, this time both of us on bikes. Down the incline from our house to the Rivers Trail, past Pioneer Park – its beach not yet rowdy with its afternoon complement of teenagers and dogs, through the flower-basketed elegance of Riverside Park, and then over the broad span of Thompson River via the Overlanders Bridge, before corkscrewing beneath the bridge, past a grizzled man pushing two grocery carts, up alongside fast-moving traffic, and then a right and a left, and we were at the farm. En route to glean cherries from a house further to the west, Maggie showed me where to lock my bike before pedalling off. She'd be done at noon, she said, and would check back with me before heading home.

I stand up from drawing and look around. At the other end of the farm, Kevin is talking with an elderly couple. Their conversation floats toward me over rows of broccoli already going to seed, heirloom tomatoes dotted with flowers, and cabbage weighty with leaves.

Closer, a middle-aged woman advances toward me.

"Hi, I'm Joanne." She's another one of the volunteers, she says, with a few beds in the back quarter. She used to farm, she says, west of Kamloops, but her ex-husband got the farm, and now she lives here in town on the North Shore. Like Sandra, she is recently retired – from Kamloops Immigrant Services – but she stops by Butler every day.

This flood of information demands reciprocity. I tell Joanne I'm a botanist from the university working to illuminate the plants in our lives – those we cultivate, those we pull as weeds and even those we ignore. That I think re-storying our relationship with plants – those quiet beings we so often take for granted – might help restore our relationship with the land and its original inhabitants.

In a plot where the relationship between plants and people is so abundantly embodied in green abundance, my words feel obvious, and I'm grateful when we're interrupted by Caitlin, the assistant farm manager, who tells me she's best friends with a former research student who worked with me. Caitlin tells me that, like Kevin, she finished a degree in political science, but working on the farm has made her learn more ecology. In the distance, a man on a bicycle rides by and waves.

"Isn't that Drew?" asks Joanne.

"Yup," replies Caitlin.

Joanne and Drew, I'm told, are farm neighbours, their personal rows abutting one another, but they haven't seen each other since spring.

I've been on the farm less than 45 minutes, and already I feel like I'm awash in relationships I don't understand. Kevin is still talking to the elderly couple, so I make my excuses and move to draw spinach plants setting seed.

Light curves around stem and leaf; in the axils, ovaries swell with endosperm and embryo.

I've moved on to the lacinato kale when I feel a lean presence above me. It's Kevin.

"I don't want to disturb you," he says, "but when's a good time for a tour?"

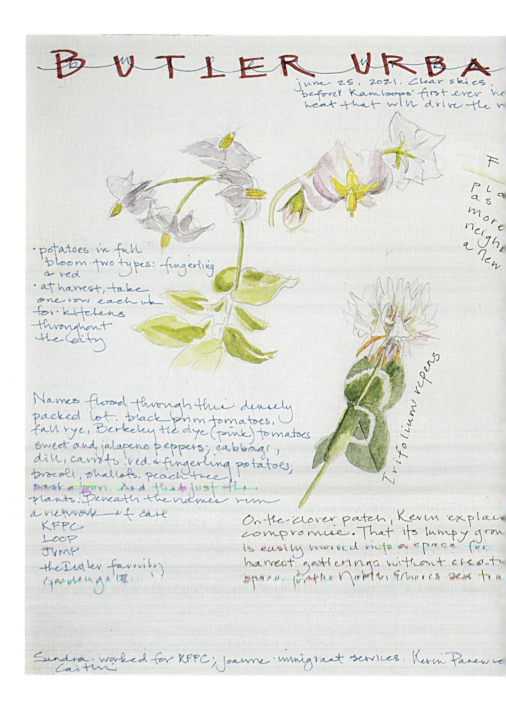

BUTLER URBA

june 25, 2021. Clear skies
before Kamloops' first ever he
heat that will drive the w

F
Pla
as
more
neighe
a new

• potatoes in full
bloom two types: fingerling
& red
• at harvest, take
one row each wk
for kitchens
throughout
the city

Names flood through this densely
packed lot: black prim tomatoes,
fall rye, Berkeley tie dye (pink) tomatoes
sweet and jalapeno peppers; cabbage,
dill, carrots, red & fingerling potatoes;
brocoli, shallots, peach tree
~~and that just the~~
plants. Beneath the names run
a network of care
KFPC
LOOP
JUMP
the Butler family
Garden gate

Trifolium repens

On the clover patch, Kevin explai
compromise. That its lumpy grou
is easily moved into a space for
harvest gatherings without creat
space for the Native Grasses sees t

Sandra worked for KFPC; Joanne immigrant services. Kevin Panewie
Caitlin

FARM

the waning hours
six days of unprecedented
3°C in three days.

lacinato kale · cavola nero · dinosaur kale
bumpy-leaved kale, leaves nearly hiding
their photosynthetic capacity. cool and elegant
in the bright light

R
Y
E

er crop, now envisioned
local graft beer, one
to reach out to its
know grown to heal

In the sunlight, the spinach gleams sap green behind
the purple green of lacinato keel; in leaf axils ovaries
swell with embryo and endosperm. Seeds grown in place.
with the hope that with each successful fertilization
will imprint the ecology of their garden on the
seeds of next years growth.

"No," I respond. "When's good for you? It's your schedule that matters."

We go back and forth until decisiveness from me feels like a mercy.

"Let's talk now," I say.

I'd forgotten this. That Kevin is a young man whose consideration of the privilege and inequity that can run, unacknowledged, in even the simplest of interactions has given him a carefulness that verges on awkwardness. In this garden, he is the expert, but over the next few hours he will check himself for mansplaining, and he says little about the farm without acknowledging the work of others. But once Kevin starts, the information comes as a flood.

We are guided by the farm's rows, but along with the names of plants – black prim tomatoes, fall rye, sweet and jalapeno peppers, cabbage, broccoli, shallots – comes a larger story. A story that comforts.

From the moment Kevin starts talking, I understand that the community supporting this farm reaches well beyond its borders. The farm, Kevin says, was for many years a vacant lot used as a druggie hangout. Even now, they still unearth IV needles. The land itself is owned by the Butler auto dealership that abuts the farm on its southern side, but it was a network of care – embodied in those groups whose names rattle off Kevin's tongue: JUMP, the LOOP, Gardengate, KFPC – that fertilized this farm into being. Later, I will sort which of these groups glean Kamloops's neighbourhoods for fruit, which use gardening to help stabilize the lives of my city's most marginalized people, which strategize how to make Kamloops more food secure. Later, I will find Caitlin's academic paper in which she interprets Butler Farm as an experiment in growing food as *commons* rather than *commodity*.[6] But, in the moment, with Kevin moving on, it's already clear how this farm's network perceives plants and people as equally potent parts of one whole.

The first summer, in 2017, he worked in the garden, Kevin says, was a season of despair. When BC's fire season began far too early in late June, not early August and the smoke settled in for the summer, the farm lost its volunteers. Without people, Kevin says, there was little to harvest. When he left the farm at the end of the season, he didn't think he'd be back. For the next two summers, weeds clawed at the farm. But then, in spring 2020, when COVID-19 heightened Kamloops's food insecurity, the farm's network reached out and brought him back. Last year, the farm grew 6,400 pounds of food. Sixty-four hundred pounds of organic produce, the tangible weight of a watershed's gifts, fed into kitchens across the city.

But this was only possible, says Kevin, because Rob at Gardengate coached him to cultivate a social network along with the farm's crops. The farm dynamic changed, he says, when they decided to set aside the back quarter for private beds. Unlike the rest of the farm's common area, these rows cultivate volunteer stability. In this farm, the above ground volunteer community matters as much as the below ground microbiome. Likewise, Kevin says, one of his most important jobs as farm manager is to grow the farm into a good neighbour. The chalkboard near the driveway provides daily hours and contact information. The front middle rows of the farm are "picky rows," filled with a selection of crops so new visitors will always find something ready to pick. No one, says Kevin, should ever leave the farm empty-handed.

Within the farm's commons, most rows are mixed, but Kevin explains their planting balances the ecology of individual crops with the expertise of its weeders. Volunteers need to be able to distinguish seedlings growing together. As we walk the rows, a few plants are tagged with orange flagging. Seed plants, Kevin says. And then he's off – describing the testing of plants within place that is the foundation of any good farm.

Some new varieties come from afar; many more come locally. Along the fence are the young shoots of garlic grown from bulbils harvested from Kevin's parents' garden. The next row is a wash of bright magenta leaves. A volunteer family, recently emigrated from the Philippines, he says, has been teaching him to grow amaranth for leaves, rather than for seeds. Elegant stalks of fall rye border the chain-link fence. The rye, he says, they planted as a cover crop, but they couldn't resist letting it go to seed. Now the farm is conspiring with a neighbour in the nearby trailer park who brews beer.

Abruptly, there's a shout from the back of the garden. We turn, and it's another man whose profile I recognize. A man who sits on the board of the farm, who helped our local naturalist club secure the Canada Jobs grant that's been employing Maggie all summer. Kevin excuses himself, and I welcome the chance to linger alongside the rye.

Secale cereale. Each year, I teach the history of this plant's changing relationship with humans that textbooks call *domestication*. I know the redistribution of this plant across the world, along with many others, helped set in motion the industrialization that spewed carbon into the atmosphere, the European colonization that planted the orange- and red-shirted crosses standing on the highway. I teach this species, but I rarely get to draw its flowers. As my pencil finds page, I fall into this flower's rounded repetition of glume and lemma, its linearity of awn and stem. Hints of quinacridone gold slide into the glumes, but the lemmas are still sap green, fixing carbon, sucking down the world's hurt. As a species, *Secale cereale* might be exotic to the Thompson River watershed, but it's also surprisingly good at growing on marginal, degraded sites. Before he left, Kevin said this row near the fence was one of the worst on the farm.

WHEN KEVIN RETURNS, I've wandered to the back of the farm, near the compost bins in the corner. Donated, Kevin says, one from a volunteer, one from the regional district. But it's only when I turn around to view the full reach of rows stretching toward the chalkboard, the small tool shed at its front, that I understand the full revolution of this farm.

There are no gates at Butler.

Such a simple statement. Such a complex reordering of the modern world's normal parameters. Chain-link fences separate its rows from its neighbours to the east and west, the terrace of asparagus and peach and Saskatoon limits access from the south, but to the north, the farm is completely open to the street.

No fence, no gate, merely a chalkboard.

"You can't steal food from Butler," Kevin says beside me. "You can only harvest incorrectly."

"If this thing gets fenced," he continues, "I don't want to be a part of it."

WE ARE IN THE NEWEST PART of the farm – a triangle adjacent to the steep slope holding the street above our heads. Until recently, the triangle stored compost and mulch. Now it's filled with herbs and flowers and more vegetable beds. I'm still trying to wrap my head around the farm's lack of gates as Kevin points out rows of basil, fava beans, chickpeas. It's the first time I've ever seen chickpeas in the ground, and I want to fix their green form in my mind, but Kevin has moved on to the pollinator garden.

"It was a compromise," he says.

Kevin explains further, "I wanted a space for gatherings. I thought it'd be good to have a space where we could celebrate the harvest. But the neighbours worried that if we planted a lawn, then it would become a space for prostitutes to bring their johns."

I'm startled by Kevin's matter-of-fact incorporation of North Kamloops's sex trade into the farm's design when Joanne interrupts. She's finished, she says, and wants Kevin to know she's leaving for the day.

Kevin notices her empty hands. Doesn't she want some baby carrots? Caitlin just finished thinning. "Here, let me get you a container that we can fill and weigh before you go."

In the relative quiet, I try, half-distracted by a bee (*Megachile perihita*?) visiting the clover flowers, to understand why my initial surprise is now reverberating with second-hand satisfaction. It's not just the simple acceptance of the North Shore's social reality that I heard in Kevin's words. No, what I'm relishing is how Kevin's understanding of *both* plants and peoples – as dynamic entities with agency and behaviour – has allowed him to negotiate the competing demands of the farm's neighbours. I know that few relationships rooted alongside plants grow without tension. Yet, in its lumpy growth, clover will feed the bees, fix nitrogen and reassure the neighbours. It will also, when the time comes, mow easily into an open space for the community gatherings that will help fertilize this farm into continuity.

IN THE BRIGHT LIGHT, Kevin is walking back toward me, but trailing behind him are more volunteers, and I already feel baked by the growing heat. Just one more question. Earlier, I'd been nosy enough to ask Kevin if he thought the farm could keep him busy for the next 20 years. Maybe not 20, he'd replied, but definitely the next five.

Kevin, I ask, this fall will you come give a talk to students in my botany course? Maybe, I say, together, you and I can channel some of its students into more volunteers for the farm.

When Kevin agrees, I can't tell who's more pleased – me or him. Seconds later, Maggie pedals into view on the street above us.

I'll be back, I tell Kevin. There's more I want to know, but I've taken up enough time this morning.

Anytime, he says, before turning away to a volunteer whose name I've yet to learn.

With her container of baby carrots, Joanne walks with me up the driveway to where Maggie's waiting with her bike. "Going far?" she asks.

"Just across the bridge to the South Shore," replies Maggie.

"Ah," says Joanne. "You folks live on the *wrong* side of town."

The three of us laugh, but as Maggie and I pedal away, I can't help but wonder.

WE ARE HEADED UP the broad span of the Overlanders Bridge, my daughter once again leading the way, crossing back over the twinned waters of the North and South Thompson. I know what lies ahead will not be easy. Astride my bike, feet locked into its pedals, I fear not just the menacing heat of the building heat dome, the additional graves that will be found in communities across Canada, but the fires – in both forest and church – that are certain to follow.

I look east, toward the brick building of the Kamloops Indian Residential School, once the largest of its kind in Canada, that still stands in the heart of Tk'emlúps te Secwépemc territory. The building now contains Band offices and a daycare, but it's always a hard building for me to see. *Truth and Reconciliation.* Today, few in my community deny the cultural truth of European colonization. We've heard the residential school Survivor testimony; we know how many young Indigenous women number among BC's missing and murdered; how many of our homeless were born on unceded, undeeded traditional territories; how food insecurity haunts Indigenous kitchens far more than settler.[7] Three weeks ago, we began to learn the crushing grief of bones buried in secret, without ceremony or family. Likewise, few now deny the ecological truth that was ushered in, at least in part, through European colonization. We've seen the smoke hanging like a shroud over our valley, heard the death dates for the glaciers that feed our rivers, and now wonder how to live in a world where the past no longer predicts the future.

I think about those Europeans called the Overlanders, for whom this bridge is named. People who looked like me, settlers who arrived nearly starving and desperate, floating on rafts down the North Thompson in 1862, more than half a century after the water that carried them was given an English name. What did these people think, eating potatoes from fields emptied of their original inhabitants by smallpox? Did they notice the silence as they floated past the too-quiet *c7istktens*, the subterranean pit houses of the Secwépemc People?[8] After these people arrived in Kamloops, few stayed long.[9] I think about the people who came after, who did not hesitate to log the cottonwood forests of this river into field and garden for their seeds, who did not hesitate to understand their laws as the only laws, the story they carried – book-bound and portable, and knowing only a single god – as the only story. Did they wonder about what would grow from their world view, their insistence on separating land from body, nature from culture, wealth from ecology, even as they toiled in this floodplain?

My quadriceps are burning; we've nearly reached the height of the bridge. If the *truth* of European colonization is clear, the path to *reconciliation* is far less so. I think about the hands that have tended the plants adjacent to the Kamloops Indian Residential School in what is now called the Secwépemc Museum and Heritage Park – the hands of children torn from family, the hands of the nuns and priests that left such horrific scars, that buried children without family or ceremony; the more recent hands like Kevin's that came not to harm but to learn as he finished his diploma in horticulture. The hands that have since taken these lessons across the river to help shepherd Butler Farm into being.

Can one acre of urban cropland help reconcile the hurt of our world? I look west, down toward where the separate waters of the Thompson rivers become one. If the Anthropocene's story of separation lies at the root of today's grief, then doesn't *anything* that defies these artificial boundaries help? I think about Butler's lack of gates; its uncommodified food grown on borrowed land; its tall rows of rye. What could better exemplify reconciliation than this Old World species, much ignored and often maligned, helping to heal a New World floodplain?

Later, I will learn about the healing role gardens have played throughout North America – in cities big and small, in Japanese internment camps, in traditional territories.[10] Later, Maggie will deepen my understanding of Kevin and Caitlin's work at Butler Farm. This spring, Maggie, daughter of two botanists, theatre tech major, leveraged her diverse skills into a full-time summer job for our local naturalist group. Together with four other university students, she's been working to better integrate nature into Kamloops's urban landscapes. My daughter, quiet but strong-willed, has been, for the first time in her life, negotiating how to disagree with those who sign her paycheque. Yet, each time she's come home from Butler, she's carried nothing but praise. Did I see the shed area, behind the terrace? It's where folks still shoot heroin, and Kevin won't let anyone but him or Caitlin work back there. Too risky. Did I hear how Kevin always amends any chore assignment with "if you get bored, come get me and I'll find you something else to do."

As Maggie and I approach the south side of the Overlanders Bridge, skimming over the still-separate waters of the Thompson, I anticipate the cool, sweet glide down into Riverside Park's canopy of trees. I also find myself filled with unexpected hope. Hopeful for the shade ahead. Hopeful that, in this unpredictable world, my daughter, in learning alongside Kevin and Caitlin, is becoming adept at crossing lines – in grief or in care, with garden trowel or gleaning bucket. Maggie turns left off the bridge, sliding down into Riverside Park. I follow, unreasonably reassured that it was my daughter, this young woman pedalling away from me into adulthood, who first introduced me to the rich possibility of Butler Farm's troubled soil.

22. Reconciling  Botany Pond

IT IS JUNE 2013. Eight a.m. I am in Botany Pond again, pants wet to the knee from brushing against grasses still loaded with last night's rain, staring at the rock pile in front of me.

It's a pile of stones spit from the earth, built from rocks bigger than a cantaloupe but smaller than a pumpkin. Weighted in place by lichen thalli a century in the making. No taller than knee height, but longer than I am tall. Surrounded on two sides by a thicket of Saskatoon shrubs.

Just a pile of rocks. But out here amid Botany Pond's rolling blanket of flowers and grass, its humped mass is a discrepancy I can't reconcile. An incongruity made all the more astonishing because if I now claim to know the botany of any place, it is with *this* mosaic of open grassland, aspen groves and Douglas fir forest. Up to now, this morning had been just one more page in my apprenticeship to the place that has become the heart of home. Earlier this spring, my field botany class found chocolate lilies on the margins of the aspen stands, and I'd come back today hoping to sort out the range of this flower's extravagance: *Most abundant on the crest of hills; no, more abundant where the shrubs are few...Always in the tension zone between aspen and grassland, except where the aspen canopy is tattered and thin.*

But at the top of the slope that climbs eastward from the pond, a sharp wind had reversed my direction. And it was in that moment of turning I saw it. A rock pile that didn't belong. Or at least didn't fit with the stories I've been telling of this place and its botany.

Was this grassland plowed? It's the best explanation for this pile of sharp-edged rocks bunched together like cows against a storm. Out here, where a blanket of grass unfolds unbroken over a foundation of glacial debris, there's no other logic for their heavy presence. I grew up picking field rocks, and I know no one wants to lug them far. I contour across the slope, and just at the shady boundary of the Douglas fir forest I find several more piles – the bony edges of their rocks softened not with lichen but with heavy-pelted moss. Sitting down, binoculars up, I scan across the grassland to the pond below before shifting up to the aspen stands and fringe of Douglas fir forest on the opposite slope.

This place. Years ago, when Laura and I first came to sample its plant community, the land was owned by a ranching family in town. Now it's stewarded by Nature Conservancy of Canada. On my office bookshelf, data collected from this upper grassland are gathered into at least one master's thesis, three honours theses, and too many directed studies projects to count. Regardless of ownership or research project, the vibrant botany here, seemingly free of human imprint other than cattle grazing, always beckons. All my dogs have disobeyed me to chase deer here; Maggie has swung from its trees and rolled downhill atop its grasses. I won't even try to guess how many journal pages I've filled with its botany.

Yet, right now, staring at these rocks, I'm stunned by my apparent blindness. I've walked with Ron and Marianne Ignace on the opposite side of this valley as they outlined the depressions left by pit ovens and showed botany students how to dig for balsamroot. Given the complete occupancy the Secwépemc People have with their traditional territory, I've never doubted that Botany Pond was part of the lived experience, the stories, that have rooted the Secwépemc People in place since time immemorial. I even knew there was a history of homesteading in the area, that there was a schoolhouse not far up the road. So why did I never stop to consider if Botany Pond had felt the sharp bite of the plow? Sitting here, I can't help but wonder: Have I really missed these rock piles all these years, or was it a more cultivated ignorance?

Either way, these unexpected stones – picked by hand, lifted by a back already beginning to ache, carried by legs from where they were dropped by glaciers – clearly stain this place I treasure with Canada's unreconciled past, so much like a buried beneath their weight: the declaration of a 15th-century pope who legalized European land theft from all those who recognized gods other than his own; the genocide and dispossession that stole lives and land; the enduring resistance of a people loyal to home and territory, even as others re-storied the world into one in which land could be owned by those still stranger to it.[1] What other legacy of colonialism, I wonder, have I avoided seeing?

"Homelessness," writes Ted Chamberlin, "haunts us all."[2] Some, Chamberlin says, have been made strange in their own homes by the actions of others; others are made strange by their own decisions. Although never equivalent, neither is without grief. The way home, Chamberlin suggests, whether for native or newcomer, settler or wanderer, often begins with shared stories.

IT DID FOR ME. More than 30 years ago. With the stories of a poet.

I'd been uprooted by desire. Or so I thought. Now I wonder about the assumptions that caused me – like so many western kids – to thirst for an education "back east." Either way, when I'd arrived in Vermont, Bennington College's "literary brat pack" of the mid-1980s[3] had me longing for the familiar, if not the beloved: a Douglas fir forest stubbled in clear-cut; Wrangler jeans branded with a snoose can's imprint. Had I known John Keats's poem, I might have said that, like Ruth, I stood "in tears amidst the alien corn."[4] But poetry was part of the problem. In my first-year seminar, I understood little of the poems we read, and even less of what my fellow students said. Deep in the stacks of the college library, I was tired of lives I didn't recognize. Then I saw it on the shelf – spine uncracked, never checked out, but there.

The Lady in Kicking Horse Reservoir is a slim book of poems, written by a westerner about western things. All names have power, but none speak so loudly as familiar names found far from home: Absarokee and Butte; Helena, Milltown, Victor. Sinking to the floor, I read – no, inhaled – Richard Hugo's poems about small-town Montana. Most were blunt; some were ugly. Children huddled in cold cars outside tavern walls, old men raged in forgotten towns, and names engraved on wooden headstones "got weaker each winter." But in others a person could see cattails "flash alive" in lightning, and "magpies spray from your car."[5] Hard or soft, town or country – it didn't matter. These were stories I recognized, stories that, through their very familiarity, recognized *me*.

I have no memory of the term paper I eventually wrote, but I'm not surprised to find its typewritten pages – clasped together by a now-rusting paper clip – in a box under Marc's workbench in his shop. In my hand, it's a tangible artifact of a defining moment. Long ago, in a building filled with kids from New York City and Los Angeles, none of whom knew how to pronounce *Butte* (so that it rhymes with *mute*, not *mutt*), the stories in *The Lady in Kicking Horse Reservoir* let me *see* the land that had raised me.

Today, I know I got lucky. Ecologist Wes Jackson argues "our universities now offer only one serious major: upward mobility. Little attention is paid to educating the young to return home, or to go some other place and dig in. There is no such thing as a homecoming major."[6] If luck fell my way the day I found Richard Hugo's poetry, then I was graced with even more when my botany professor introduced me to the writing of Henry David Thoreau, Barry Lopez, Wendell Berry, Annie Dillard and Gretel Ehrlich.

Far from home, these authors tutored me in an allegiance to place. Yet it's one thing to read about other people's attachments, and another to honour your own. As a young woman in a mobile world, I was far more loyal to opportunity than to the ground underfoot. Poetry may have taught me that my home had stories that mattered, but it took the science of botany to give me reason to return. First to Montana, and then to southern BC. Once I was home, drawing and botany, place and practice, came together in ways I could never have imagined, most often in *this* mosaic of forest and grassland and wetland. The summer Laura and I first came to sample its plant community, Maggie needed a car seat to ride here with me. Now, 17 years later, Laura mothers her own daughter on Vancouver Island, and Maggie can drive me.

I MAY HAVE COME HOME, but, set against the long arc of human habitation in this place, I am not native.

Today, as I sit looking down at Botany Pond, my genes still carry the weight of my ancestors' long association with portable crops and herd animals. In this corner of the continent, north of the American border but well south of the boreal forest, along the silty banks of the Thompson River, my red hair and light skin, my ability to metabolize milk and alcohol, clearly label me as an exotic. Unlike the European settlers who picked rocks from these grasslands, I don't depend on this soil for my bread. I don't have generations of stories teaching me how to live in a good way in this place. But if I don't belong to this land, where do I belong?

I look down at my open field journal and scan through the impressions of the morning. I'd been concentrating on the pattern of chocolate lilies when a red-naped sapsucker landed abruptly in the water birch next to me. The bird was only a handbreadth away. Startled by such proximity, I backed away, trying to give the bird distance, even as there were more – a ruby-crowned kinglet, a yellow-rumped warbler – in the same shrub. Across a small hollow of sedges, I squatted, binoculars up, needing to understand.

Finally, I saw it. The sapsucker had stripped the bark on the birch, opening irregular windows into the underlying phloem. Sap ran freely. For a few crystalline moments, the world had been made whole. Kinglet, warbler and sapsucker drinking in the green glory; me made participant, by the simple, yet revelatory, act of witnessing.

I AM NOT NATIVE. But, today, I have to believe that *all* people share the capacity to be shaped by place. Today, I have to believe that the bridge from exotic to naturalized can be built, at least in part, through the attention we give to the world, especially its plants.

I know I will walk Botany Pond as often as I can and learn its stories, old and new, hard and soft. I will pay attention with every tool I have. As best I can, I will lend not just my eyes, but my heart and my tongue, in support of this land.

But is it enough?

If Botany Pond's the heart of the land I call home, it also sits within the Dominion Railway belt, the 40-mile swath of land given to the Canadian Pacific Railway to subsidize the building of the railway, and to populate what is now called Canada with settlers. I know that foundations of homesteader cabins – surprisingly small depressions, overgrown with snowberry and rose – stain this landscape with a history of pre-emption. By the time the Dominion homesteaders set foot in Botany Pond, the Secwépemc People – governed by the Indian Act, corralled into reserves, constrained by travel passes that needed to be signed by an Indian agent – had already been dispossessed of their long-standing relations with their traditional territory, including Botany Pond and its plants.

Let's be clear. This dispossession was no gentle shepherding of people from one side of the valley to another. In BC, colonialism, enacted across the centuries, was a violent strategy to mobilize and transfer the wealth of the land, trap its furs, cut its trees, extract its gold, and replace its native grasses with crops from other continents. Colonialism understood the richly storied and occupied landscape of Secwepemcúlecw as empty land "awaiting development and its inhabitants as backward and lazy."[7] In BC, colonial dispossession transferred 99 per cent of the land from Indigenous to settler stewardship, and sent seven generations of

Indigenous children to residential schools. Schools in which children as young as 5 or 6 were told, in the words of Justice Murray Sinclair, chair of Canada's Truth and Reconciliation Commission, "that their lives were not as good as the lives of the non-aboriginal people of this country; they were told that their language, their culture was irrelevant; they were told that their people and their ancestors were heathens, pagans and uncivilized and they needed to give up that way of life."[8]

Imagine them: Young children, not just standing like Ruth in tears amid the alien corn, but children mentally, physically and sexually abused. Children buried in unmarked graves. Children whose loss continues to haunt family and community.

A pile of stones – 215 crosses dressed in red and orange, standing sentinel along the highway. How can any of us – native or newcomer, naturalized or exotic – ever belong if we don't find a way to acknowledge, to reconcile, these stories?

IT IS OCTOBER 2015, and I stand in the Grand Hall of my university, in a room full of Secwépemc Chiefs, in a room full of residential school Survivors, and listen, heart rising, as Justice Murray Sinclair is drummed into the room. In front of me, a small woman in a bright blue hoody – maybe 10 or 15 years older than me – rotates her hands, back and forth, back and forth, in the space in front of her heart. It is, I understand, an embodied gesture of respect. On stage, Justice Sinclair is a big man with a melodious voice. Dressed in a gray suit jacket overtop a pale blue sweater, a choker of red and white beads emerging from under his shirt collar, he stands behind the podium with easy grace. After thanking the traditional keepers of Secwepemcúĺecw, he particularly thanks the drummers for his welcome, noting that, as he has travelled the country gathering the stories of Survivors, he never knows how he will be introduced.

One time, he says, a man stood up in the room and declared in a loud voice, "Everybody, shut up. Sit down. He's here."

Tonight, we're already sitting and silent, but our laughter settles us, even as we know the next two hours are going to be hard.

It is. The truth is harder than anything I've sat through before. My only comfort comes from the willingness of three Indigenous students – one from the Cold Lake Band in Alberta, one from Haida Gwai and one from Gitksan Territory – to sit beside me. All have taken my classes. I've walked with two in the field. One, Jordan, came with me to Botany Pond after reading an earlier version of the essay that now makes up this chapter.

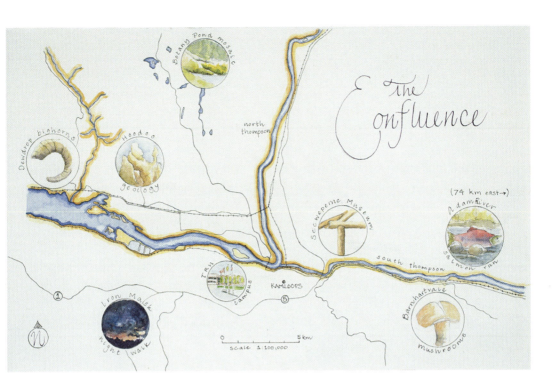

Illustrated Map: The Confluence

It was near the end of the semester. In a new "natural history" course, in which I'd asked students not to memorize facts but to consider "the lessons a ponderosa pine might teach you." I'd introduced the students to field journalling, and then as a group we'd chased butterflies along the Tranquille River, illustrated artistic maps of *home*, night-walked under cold, star-studded skies, visited with spawning sockeye salmon and rutting bighorn sheep, and, most importantly, been guided through the exhibits at the Secwépemc Museum. I hadn't known enough then to ask if Jordan's parents were Survivors, and had been horrified by my ignorance when Jordan abruptly left the museum. Later, he told me he couldn't breathe in front of the photos of the children.

On stage, Justice Sinclair, no stranger to the power of rhythmic language, says, "We have heard the truth and talked the truth." When he goes on to state, "But it is no longer about you," the woman in blue sitting in front of me nods her head vigorously. Later in the night, I will learn her name. Later in the night, I will learn that she, too, is a residential school Survivor, but in the moment, I am already humbled by her willingness to look forward.

Justice Sinclair continues, "If you thought truth was hard, reconciliation will be even harder."

IT MIGHT BE HARD, but it is *our obligation*. If the words of a poet first pointed me home, then it was the words of Mike Arnouse, the Secwépemc Elder I sat across from during the Blanket Exercise, a ceremony of truth and reconciliation earlier today, that let me, perhaps for the first time, understand the responsibility that comes with returning home.

Uncle Mike's a familiar figure on my university campus. He gave the blessing for tonight, as he has for many official gatherings, first in his own language Secwepemcstín, and then in English. I know Jordan trusts and respects Uncle Mike, whom some say "moves like a mountain." During the Blanket Exercise, Uncle Mike sat perched forward in his chair, three-pronged cane under his broad hand. We were in a room across campus from tonight's Grand Hall, high up in a building that gave us a direct sightline across the Thompson River to where the grasslands of Botany Pond roll up to meet the treeline.

We stood on a cacophony of brightly coloured quilts – a visual riff of intersecting patterns – each one representing the territory of one of Canada's original First Nations. The orchestrators of the exercise moved us figuratively through the truth of Native–newcomer history. People died (and had to sit down) as smallpox decimated whole villages, or were lost to alcoholism or to residential schools. People crowded together on fewer and fewer quilts as more and more land was pre-empted by settlers. A startling few gained the space of an unfolded corner in recognition of several recent Supreme Court of Canada decisions. The government – represented by a nursing instructor dressed in the top half of a bright red military uniform – relentlessly patrolled our outer limits, pushing us onto the tiny scraps of cloth representing the 1 per cent of British Columbia now delimited in First Nations' reserves.

It was not until we sat back in our chairs and our time was nearly gone that Uncle Mike spoke. Many in the room – those whose lands had been stolen, those who had lost parents to the trauma of residential schools, those who were beginning to understand the responsibilities that come with inheriting stolen goods – had already spoken. Mike's words were short, maybe even a little abrupt, with anything that smelled of hesitation.

The land, he said, needs us. It was not long ago that the Adams River salmon run was twice as large. And when he was out last, sitting on the bank watching the salmon, he grew concerned with the activity of the white-water kayakers. The salmon wouldn't go up, he said, with the plastic boats splashing about. But when he waved the kayakers over, one of them got mad, telling Uncle Mike he had as much right to this land as anyone.

Then why, asked Uncle Mike, "am I the only one looking after the river?"

WHAT IS MY DUTY TO PLACE, to both its plants and its peoples, as a child of newcomers? How can any of us – those who were born on this continent, or those who are arriving today – transition from being displaced Europeans or Asians or Africans to become inhabitants of this continent, this watershed?

The land needs us, Mike says. But what does it need most?

I need to be careful here. If re-storying our relationship with plants is as key to surviving the Anthropocene as I believe it to be, then our attention may well be the greatest gift we have to give. But not just any kind of attention. If we are to make kin with the green inhabitants of this world, our attention – cultivated through gardening or drawing or gathering – needs to weave our story in with both theirs and those who have known them the longest.

Today, the First Stories of plants and place must be fundamental to my understanding of home. They are not mine, but I will learn from them. For this reason, Ron and Marianne Ignace's *Secwépemc People, Land, and Laws* sits on my bookshelf next to Thoreau's *Walden*, the reason that Art Manuel's *Unsettling Canada* sits next to Wendell Berry's *The Unsettling of America*, Garry Gottfriedson's *Skin Like Mine* sits next to Richard Hugo's *The Lady in Kicking Horse Reservoir*. It's important to know the plant I call *Lomatium macrocarpum* is both an important food root of the Secwépemc People and a mythical being called Qweq'wíle. And that the story of Qweq'wíle, with a few exceptions, circumscribes both the range of *Lomatium macrocarpum* and the "boundaries of what would become Interior Salish territory, with Secwépemc territory at its core."[9] It's important to know that the land that has helped raise my daughter has been protected and defended by the Secwépemc People since time immemorial. That, collectively, the people of this territory have resisted, and continue to resist, colonialism – under the cover of night, on roadside blockades, on scarred pipelines corridors, within courtrooms, in front of the United Nations.[10]

Learning *about* and *from* the plants and people of place is not the work of a week, or even a year. Nearly two decades ago, I arrived home and wondered what it would take to be as rooted as a tree. Cottonwood, *Sphagnum*, spotted knapweed, *Crossidium seriatum*, sagebrush buttercup, licorice mint, lichen, fall rye – each has taught me. As plants, they speak of soil and elevation, of lives known or not by humans. As neighbours, and sometimes even kin, these same species have rarely let me rest easy in my assumptions. Together, plants and place have challenged me to answer the question: *How do I belong?*

It's not an easy question, but for someone like me – a descendant of settlers, a child of back-to-landers, a woman who has used a colonial science to feed her family – it might be the most important question I will ever answer.

IT IS LATE AUGUST 2021. One more morning in Botany Pond – this time I'm with three schoolteachers from an outdoor school in the North Okanagan. Earlier this spring, I gave a field journalling workshop for their faculty. Now three of them have asked to spend more time with me in the field. Of course, I suggested we go to Botany Pond.

None of these teachers lack experience with the living world; all of them are as devoted to place as I am. Yet, collectively, all four of us are still reeling from the overlapping crises of this summer: the horrifying announcement of the unmarked graves on the grounds of the Kamloops Indian Residential School, the terrifying heat dome of late June, the fires of July and August that evacuated tens of thousands of our neighbours, the smoke that extinguished summer with worry, the latest COVID surge that has limited all nonessential travel to our health region.

But last week the smoke cleared and, today, calm feels possible. We walk the upper trail to the pond, contouring through the aspen stands, past the open-grown Douglas fir tree with branches nearly to the ground, and I guide the teachers past the smaller wetland where I've learned to avoid the sandhill cranes (*same year after year?*) who nest nearby. Today, the cranes are busy elsewhere, but when we hear them bugling in the distance, all of us stop, mid-stride, to listen.

It is when we find the wild onions dropping seeds that our conversation turns from place to mobility. To the cultural patterns of mobility found in settler culture that have reshaped the world. As individuals, all four of us have used mobility for opportunity; this summer, none of us have been able to avoid its temptation as escape. Since I saw them in April, the teachers, like me, have been see-sawing between grief and rage. As I pull out paper bags to collect onion seeds for the new pollinator garden on campus, I tell the teachers this is my first trip to Botany Pond since spring. That, in early summer, when less than 20 per cent of our normal rain fell, fear of what I might find kept me away. That, in July, when the record-breaking heat and smoke trapped me, Marc and Maggie like rats in our basement, I began to wonder how many more summers in Kamloops – enraged by fire, shrouded by smoke – I could tolerate. If, five years ago, some thought climate change might diversify the agricultural opportunities of southern BC, we now understand how quickly, at 47.3°C, whole towns burn to the ground.

The very fact I've considered abandoning the land that has supported my family and helped raise Maggie feels deeply shameful, but the teachers tell me I'm not alone. They are as worried as me; they, too, have considered where else they might go. Today might comfort with its clouds and cooler temperatures, but we all know what's coming. We know the future is hotter and more unpredictable; we know the consequences will extend far beyond our species. I can't decide what to worry about most: our ever-growing risk of fire, or our lack of winter snowpack. Since time immemorial, it has been the meltwater of glaciers that has kept this watershed's rivers flowing in late summer. What hope will the Adams River sockeye have when the glaciers no longer provide late summer melt in which to swim?

But then the path winds us out to the narrow bluff where I look for the purple of shrubby penstemon each May. Below us, grasses roll down to aspen patches, to the mirrored sky of Botany Pond, and then back up to the conifers beyond. It's a view that, like the bugle of the sandhill cranes, stops us in our tracks.

In the silence that follows, I look down to where I know rock piles have hunkered in place for more than 100 years, and then west to where I found the homesteader cabin foundation just uphill from Botany Pond. Did the hands that picked those rocks, that cut trees for cabin walls, worry about how to belong? Did they miss the lands from which they came, or were they just grateful for the opportunity this seemingly empty land offered? As the years unfolded, and their children came close to starving, as it became clear that their homesteads would fail,[11] that the climate of these grasslands could not support dryland farming, did they wonder about the stories that had brought them here, the stories that would let them leave, the stories that, if they had learned them, might have let them stay?

Out here, in air temporarily cleared of smoke, I know the story I learned first – the story that gave me an easy mobility, a profession, a reason to come home; the settler story that for too long we've assumed is the *only* story – has failed. Failed the original inhabitants of this territory, failed the land, failed us.

Looking down across the tawny grasses of Botany Pond, it's clear. We perpetuate this story at our own peril. Ignorance is no excuse; *this* is the story we tell each time we walk unseeing past a pile of heaped stones; each time we assume that lessons learned from afar need no translation; each time we rest easy in our occupation of stolen lands; each time we look to escape the land's vulnerability. The world is grief-stricken and imperilled, but running away will only make it worse. Our only hope, in the words of one writer, will be if we find the courage to "stay with the trouble."[12]

As we start down the hill toward Botany Pond, its waters shimmering blue, I wonder if part of the answer of how to *belong* isn't written in the word itself. Would the world be in so much trouble if more of us saw ourselves not as homesick immigrants standing in alien corn, but as inhabitants loyal to place? If we truly relinquished the privilege that comes with mobility?

To *belong* is to *be long*. It doesn't sound like much, but even just thinking the words makes me hesitate – especially now, when so much of what I have come to know and love in Botany Pond, and throughout the Thompson watershed, feels imperilled. *Stay? Forever?*

It's not a story many of us, including me, know how to tell well.

To tarry, to dwell, to stay put. Not to sleep with your go-bag by your bed, but to act as though you always have been, and always will be, a part of *where you are*, while still honouring the rights of those who came first. To *be long* is to understand that your well-being always has been, always will be, tied to the ecosystems and the people that surround you. To *be long* is to know the plants and the poets and the Elders of your watershed as well as, if not better, than those from away. To *be long* is to know in your bones there is no escape. To *be long* is to do everything you can today, so that long after you are gone, your daughter and her daughter's daughters will be able to walk through Botany Pond just as the chocolate lilies bloom and the red-naped sapsuckers glean from the living heart of birch.

To *be long* – it may be hard, it may take all I have, but it may be the only story that will save us, people and plants, together in place.

Form Follows Function

I'M IN THE HALL. Banished from my normal place inside the botany laboratory immediately behind me. Across from me, brown brick walls run left and right, punctuated at distinct intervals by glass cabinets housing biological artifacts. The thick wash of bodies moving past subsides to a few isolated quick walks of students nearly late, and then echoes into an awkward silence. Sitting here, momentarily immobile in a space clearly designed for travel, I realize for the first time how these brick walls have constrained my working life. More trips made through this hall, to this door, than perhaps to anywhere else on campus.

Years ago, when I arrived in this building, I found its structure disorienting. All the halls looked the same, and even with just three floors, I got confused between the second and third. They'd put me in the basement that first December, and, coming in early, going home late, I'd scurried from office to laboratory. I'd missed my very first Science Faculty Council, confused about which events a brand new, probational faculty member was to attend. Surely, I thought, one was *elected* to faculty council. I'd gone instead to the botany lab to practise cutting geranium stems into one-cell-thick sections with a razor. Following the directions left behind by my predecessor, I dyed individual sections with toluidine blue stain, and mounted each on a microscope slide. I remember gasping the first time I saw the crystalline beauty that lies at the heart of plant biology.

Cells with different functions – photosynthesis versus support, protection versus conduction – have walls that differ in thickness and construction materials. In a fresh stem brimming with chlorophyll, these different cells mass together in an undifferentiated green. Dark blue in the bottle, the magic of Toluiduine Blue is that it colour-codes cells by the type of molecules that construct them. A 30-second soak in this blue liquid turns plant tissue sections into stained glass

RAWING BOTA

Colours in the Landscape

Early morning light, Checkerspot butterfly wing, Dried balsamroot petal
HANSA YELLOW QUINACRIDONE GOLD

Hazelnut bark, Male cottonwood flower, new Sarsaparilla leaves
QUINACRIDONE GOLD CARMINE

Twinflower petal, Cloud streak at sunset, Decomposing cottonwood leaves
POTTERS PINK RHODONITE GENUINE

Douglas fir canopy, Shadows in the grassland, Dry hill in winter cold
SAP GREEN/COBALT BLUE BURNT UMBER

Sagebrush leaf, Glacial flour filled lake water, Pale gentian petal
CERULEAN BLUE CADMIUM YELLOW

Fresh aspen leaf, Distant hills in evening light, Subalpine turf
GREEN GOLD RAW SIENNA

Aspen bark, Sun-filled spruce canopy, _Fucus_ blade at low tide
COBALT BLUE QUINACRIDONE GOLD

Cloud bottom in a storm-filled sky, Woven grass in a flycatcher nest
COBALT BLUE BURNT UMBER

Arbutus leaf about to fall, Castilleja bract, Larkspur petal
CARMINE FRENCH ULTRAMARINE

Coralroot flower, Deepest shadow in conifer canopy, Floating loon feather
QUINACRIDONE VIOLET PHTHALO GREEN

Sun-filled sky, Harebell petal in subalpine meadow
CERULEAN BLUE

Lupine petal, Cold shadow in white snow, Lazuli bunting plumage
COBALT BLUE

THE TEN ESS

BINOCULARS

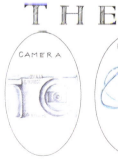
CAMERA

HAND LENS

PENS and PENCILS

FIELD JOURNAL

BRU

Location Map for Drawing Botany Home

HOME

Places in the Landscape

1. MURTLE LAKE
   Thompson Watershed, BC
2. PLACID LAKE
   Thompson Watershed, BC
3. CLEARWATER VALLEY
   Thompson Watershed, BC
4. HEFFLEY LAKE
   Thompson Watershed, BC
5. ADAMS RIVER
   Thompson Watershed, BC
6. BOTANY POND
   Thompson Watershed, BC
7. KAMLOOPS
   Thompson Watershed, BC
8. POLLY'S COVE
   Sackville Watershed, NS
9. DESOLATION SOUND
   Salish Sea, BC
10. MALASPINA PENINSULA
    Sunshine Coast, BC
11. ARMSTRONG
    Okanagan Watershed, BC
12. IRISH CREEK
    Okanagan Watershed, BC
13. THREE-MILE POINT
    Okanagan Watershed, BC
14. LINCOLN COUNTY
    Kootenay Watershed, MT
15. LOPEZ ISLAND
    Salish Sea, WA
16. HWY 97
    Okanagan Watershed, WA
17. HELENA
    Missouri Watershed, MT
18. YELLOWSTONE LAKE
    Yellowstone Watershed, MT
19. MILL CREEK
    Klamath Watershed, CA

TIALS

WATER BOTTLE

HAT and GLASSES

BACKPACK

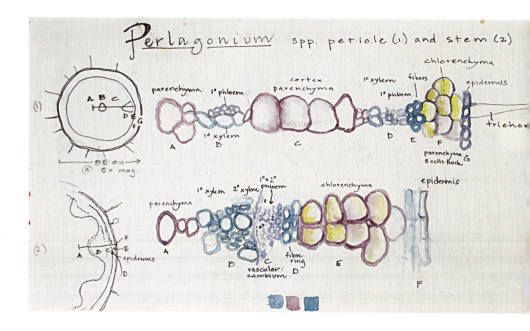

Vol. 37: Perlagonium and Tradescantia

windows of green, purple, pink and turquoise blue. Along with differences in cell shape, the differential staining of Tol Blue allows botanist and student alike to decipher the logic by which molecules build into cells, cells into tissues, tissues into stem and leaf and flower.

Inside the lab, students in my natural history class are completing a mandatory course evaluation. In a few minutes I will be allowed back in for the culminating exercise of this course – not a typical end-of-term lab exam but a writing workshop. This past Monday, in groups of four, these students from departments across campus (theatre, general studies, interdisciplinary studies, history, natural resource science, and biology) submitted essays to each other. It is now Thursday 10:00 a.m. Over the last two days, they were meant to read and comment on each other's drafts. In a few minutes, I will go in. Not to lecture, not to discuss, but to listen. Above all else, my job will be to mask my own anxiety. I want this to work. I want the students to be supportive and constructive. I want them to engage with each other's writing, to hear the new stories they are trying to tell. It's just not the type of teaching I ever expected to attempt in this botany lab.

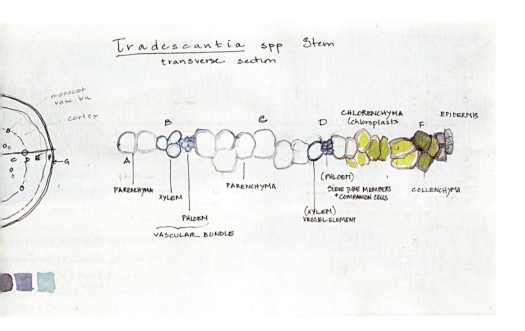

Tradescantia spp. Stem
transverse section

monocot vasc. bu.

cortex

B

e

D CHLORENCHYMA (chloroplasts) EPIDERMIS

F

A

PARENCHYMA

XYLEM

PARENCHYMA

(PHLOEM)
SIEVE TUBE MEMBERS
+ COMPANION CELLS

COLLENCHYMA

PHLOEM

(XYLEM)
VESSEL ELEMENT

VASCULAR BUNDLE

Later this afternoon, the laboratory beside me will be swamped with students from my botany class sectioning and staining, analyzing and describing, in their final lab exam. Implicit in all they do will be the idea that form follows function. Biologists say it in one breath. How, we ask students, does the anatomical structure, the form, of a leaf or a stem or root reflect its function? We know there are exceptions, but when we look at the internal anatomy of any plant organ, we interpret the structure of individual cells – sclerenchyma and collenchyma, palisade and spongy parenchyma, tracheid and vessel element – as evolutionary adaptations that allow them to function. Complicit with this reasoning is the understanding that evolution doesn't innovate out of thin air. Evolution, we assert, acting upon the variation that arises from chance, carves what works from what doesn't.

Sitting here, waiting to go in, I wonder if the same doesn't hold true in teaching. Just as no flower appears wholesale, newly minted on the face of this earth, no course appears out of thin air. Maybe the anxiety I feel, the risks I nearly wish I hadn't taken, are a necessary part of carving what works from what doesn't. I'm sure the students must be almost finished. I take a deep breath and, as the door beside me opens, find some small comfort in the familiar geography of the room I have known for nearly two decades.

Inside the lab, it's bright. Students mill at the back bench, filling mugs with tea. Earlier, in celebration of our last meeting together, I stacked bakery scones next to the teakettle. Collectively, our group is diverse. Unlike in most of my courses, I am not the oldest person in the room; neither am I the only parent. Three of us have given birth, two parent daily and one waits for her first grandchild. We include the descendants of European settlers, an Inuit man from northern Newfoundland and several whose parents were born in Southeast Asia. Two of us are terrified of birds, at least three fish on a regular basis and one has hunted polar bear. One young woman grew up in a motel room; nearly half have never lived anywhere else but Kamloops. In the room, the students' energy verges on manic; some, I think, are sleepless. Or maybe just nervous. Certainly, everyone feels the imminent release of the end of term.

This is the second essay the students have workshopped together. They are in different groups from the first workshop, but these students now have a wealth of experience with one another. We've learned from mapmakers and fisheries biologists, astronomers and ornithologists. It hasn't all been easy. I lost the two self-proclaimed theatre girls in the grasslands, and was given a stern verbal reprimand (with the threat of a $1,000 fine) by a sanctimonious railway cop when we traversed the tracks in an unregulated crossing to walk beneath hoodoos. Our visit to the residential school exhibit at the Secwépemc Museum and Heritage Park left no one unchanged.

Above all else we've read: Ted Chamberlin and Candace Savage; Kathleen Jamie and Robert Macfarlane; Wayne Grady and Dick Cannings; Keith Basso and Barry Lopez. All writers chosen because they had something to say about the way place had shaped their lives. In the first version of this course, I'd worried about the presumption of a settler like me *teaching* the works of Indigenous authors like Richard Wagamese, Jeannette Armstrong and Garry Gottfriedson — authors whose language was rooted in the very land we walked. But Richard Wagamese writes, "Stories are meant to heal."[1] By the second offering of this course, I'd understood.

If I wanted my botany, in the form of this course, to re-story relationships, we – the students and I – needed to read these authors together, to struggle with and learn from their lessons as a collective group. Certainly, a hesitation to hear

Indigenous voices was never part of the students' response to the residential school exhibit. A recognition of its horror and shame, *yes*; a desire to know more, to do better, *yes*; hesitation, *no*.

No writer has gone without discussion. In class, some, including me, have talked too much; others have rarely spoken up. I've learned to trust what the students say in their field journals far more than what they say in class. In many ways, the students have remained true to their original disparities – the theatre girls nearly always want more character development, the biology crowd more facts, but none question the common ground of the course. That is, the stories we tell matter. As Tom King argues, if we want a different ethic, we need to tell a different story.[2]

When everyone has their tea and scones, the groups separate to the four corners of the room. I look down at the stack of essays on the lab bench. I know the topics are diverse, many extending far beyond botany. I also know it's not just me who has taken risks. Woven in among stories from the land are other stories: a student's nearly instinctive desire to please her father; the weight of unwanted burdens; a fear of working dead-end jobs forever; the hope of new beginnings; the loss of those they've loved; the grief of living in an unpredictable world.

In class, we've talked about the power of constructive feedback, about what it means to support each other as authors, about how to separate the narrator from the author of each piece. From my position behind the black lab bench at the front of the room, the fume hood throbs. It's a constant irritant in this room, but for once I am glad for the white noise that separates one group from another. I slide to one end of the front bench and listen to the closest group.

"I love your similes and alliterations, but I got bogged down in the science in the second paragraph."

"I think there's an opportunity to build up the character of the narrator's father, to set the scene of the phone call when they're talking about navigating by stars."

I slide to the other end of the bench as the conversations continue – some similar to the words I penned earlier on each essay; others strikingly divergent. Not all the students remember to remain quiet as feedback is offered – the lure of ideas pulling them across established protocol. But what thrills me is how the feedback goes far beyond mere grammar. Responses tackle structure, point of view, narrative arc, meaning. The noise in the room builds; in multiple groups, laughter erupts.

I look up; with its immobile lab benches, this laboratory was built for dissections, not writing workshops. The only space for small groups to gather is in the four corners. With two groups clustered at either end of the metre-tall front bench, I'm penned in behind. But I need to hear more. Not wanting to disturb either group, I hoist myself up, standing tall atop the bench, seeing the room from a new perspective, before slipping back to the floor on the other side.

In the next group, one woman submitted her essay late – only last night – and her group is firm yet generous.

"You missed the deadline, so I didn't have time to make comments, just to read it through once."

"If you want, I'll finish reading it and send comments to you."

In each group, I watch students write down their peers' comments, trusting the guidance of one another. Not all will be incorporated. Based on our last workshop, I know some will prioritize what I say over that of their colleagues in their final edit.

Do these students understand the gifts they give each other? Not just the specific words or insights, but the gift of themselves as readers. The opportunity they provide to be heard, to be considered.

I will, after the final drafts are turned in, assess the written comments they gave one another. But I have neither the mechanism nor the need to mark what is happening right now. What I see in this room – in this one cell nested within the larger geometry of an institutional building – is a generosity of spirit that is far less about grades and far more about the accountability of community. I know the ability to rethink my story was transformative: How could I not want the same for these students?

Nearly in unison, each group is working on their last essay. As the final group wraps up, one student asks that I create another Dropbox folder for sharing the final essays. She'd like to read everybody's.

The workshop is over. But the students linger. Before they go, they get me to agree to teach a bookbinding workshop over the Christmas break. They want to hand-bind their next field journals like mine, rather than rely on store-bought ones. They'll organize it, they say. You give us the list of materials, and we'll get everything ready. You'll just need to show up.

It's time. I shoo the last student out the lab door.

Form follows function. Does the same hold true in teaching? Maybe. And maybe this slow-growing realization is the reason I'm still vibrating with satisfaction, even when all the students have gone. After nearly two decades of university teaching, this course is the one where I've had the most freedom to prioritize *doing* over *knowing, reflection* over *memorization*. As I wash the last of the teacups and stack them in the cupboard, I think the function of this course has rested far less in its built environment, with its immobile benches and too-loud fume hood, than in its willingness to let students learn from the land outside.

It's not been solitary work. Plants might be my most important teachers, but they are never the only teachers of place. My fellow science faculty and local naturalists have provided much of the scientific background in this course, and my writer friend Elizabeth mentored me in teaching writing. Today's workshop is not the only piece of this course that's felt risky. Earlier this semester, I found myself limiting how much I shared with a member of my own department, worried they would find the work of this course trivial. After all, the class has analyzed no statistics, compared no molecular data, dissected no bodies.

The primary tool of the course has been nothing more than the students' field journals. These books served as vessels into which the students poured their reading responses, field sketches and notes from our guest lecturers, before distilling out a weekly reflection. No architecture – whether it is of stem, laboratory, book or course – is neutral. I've not been surprised to find the students' weekly reflections growing more powerful, more potent, over the semester. Neither have I been surprised that many students used the meaning they cultivated within their field journals as the skeleton for the essays they've just finished workshopping.

BUT HERE'S THE THING: In evolution, innovations have a life of their own. Sometimes an evolutionary innovation that has been sculpted by natural selection for one function will be co-opted for another. Long before feathers lifted dinosaurs into birds, proto-feathers – originating as wiry scales on the backs of flightless dinosaurs – served as a means of keeping warm or attracting mates. Everything in this course – the field journal, the readings, the writing workshop – was set up to allow students to be influenced by the stories of the land and its peoples. But maybe before now I never understood the full reciprocity of story.

I have no doubt these students' essays, with their fine and careful attention to place, with their considered yet demanding questions, with their willingness to mix the personal with the ecological, art with science, will help their authors – native or settler, transient or resident – *belong to* whatever landscape they decide to make home. But, in doing so, won't these same stories transform the world? Won't they help turn indifference into love, avoidance into action, strangers into kin?

I can't know anything for sure until I see the students' final work, but I think back to the hubbub midway through the workshop, when one group was laughing and another was wrestling with verb tense; one with character and the other with how much science is too much. In that moment, the form of the preceding semester culminated in function. All 16 students engaged in each other's reimagining of the world. All acting with confidence in their roles as readers, as authors of stories that matter. Students finding human truths, building relationship, through the moments of a wounded, yet still momentous, world. A sight no less incandescent than the stained *Perlagonium* sections botany students will make this afternoon in my lab exam.

Do the students who just left understand the gift they've given me? That their stories are seeds packed with the radicle possibility of rooting, enlivened with photosynthetic cotyledons, rich with the attention of inexhaustible meristems? That their stories answer the question I asked nearly two decades ago: *What would it take to be as rooted as a tree?*

Few trees root alone. If stories link people and place, don't all of us root best when we understand our stories are but one stem among many? When we have reason to believe that our stories – of grassland or garden, bog or forest, city block or streamside margin – will continue long after we are gone?

I push open the lab door to leave. This Thursday there's no faculty council. I turn left, heading for my office, navigating without thought through the brick lined hallways of the place where I work. Later this afternoon, as the sun seeps from a winter sky, I will walk home to my family's bright blue bungalow on Pine Street, not far from where cottonwood trees braid twig into branch over the confluence of two big rivers. As always, my field journal will come with me.

— NOTES —

The Comfort of Buttercups

1 Natasha Myers, "From Edenic Apocalypse to Gardens against Eden: Plants and People in and after the Anthropocene," in *Infrastructure, Environment, and Life in the Anthropocene*, ed. Kregg Hetherington (Durham, NC: Duke University Press, 2019); Yi-fu Tuan, *Space and Place: The Perspective of Experience* (Minneapolis: University of Minnesota Press, 1977).

2 James H. Wandersee and Elisabeth E. Schussler, "Preventing Plant Blindness," *The American Biology Teacher* 61, no. 2 (1999).

3 Terry Allen Simmons, "But We Must Cultivate Our Garden: Twentieth Century Pioneering in Rural British Columbia" (PhD diss., University of Minnesota, 1979).

4 Sarah Beach, "Curse of the Hippie Parents," *Salon*, August 22, 2001, https://www.salon.com/2001/08/22/hippie_parents/.

5 Willow Yamauchi, *Adult Child of Hippies* (London: Insomniac Press, 2010), 134, 144.

6 Rebecca Kneale Gould, "Back to the Land Movements," in *Encyclopedia of Religion and Nature*, ed. Bron Taylor (London: Continuum, 2005), 148.

7 Justine Brown, *All Possible Worlds: Utopian Experiments in British Columbia* (Vancouver: New Star Books, 1995).

8 Gould, "Back to the Land Movements," 148.

9 Henry David Thoreau, *Walden and Civil Disobedience* (New York: Signet, 1989), 113.

10 F. Marina Schauffler, *Turning to Earth: Stories of Ecological Conversion* (Charlottesville: University of Virginia Press, 2003).

11 Robert Michael Pyle, *The Thunder Tree: Lessons from an Urban Wildland* (New York: Lyons Press, 1993).

12 Philippe D. Tortell, "Earth 2020: Science, Society, and Sustainability in the Anthropocene," *Proceedings of the National Academy of Sciences* 117, no. 16 (2020); Aaron M. Ellison et al., "Art/Science Collaborations: New Explorations of Ecological Systems, Values, and Their Feedbacks," *The Bulletin of the Ecological Society of America* 99, no. 2 (2018).

13 Lyn K. Baldwin, "Drawing Care: The Illustrated Journal's 'Path to Place,'" *Journal of Teaching in Travel & Tourism* 18, no. 1 (2018).

14 Yinon M. Bar-On, Rob Phillips, and Ron Milo, "The Biomass Distribution on Earth," *Proceedings of the National Academy of Sciences* 115, no. 25 (2018); Simon L. Lewis and Mark A. Maslin, *The Human Planet: How We Created the Anthropocene* (London: Pelican Books, 2018).

15 T.W. Crowther et al., "Mapping Tree Density at a Global Scale," *Nature* 525, no. 7568 (2015).

16 Francisco Sánchez-Bayo and Kris A.G. Wyckhuys, "Worldwide Decline of the Entomofauna: A Review of Its Drivers," *Biological Conservation* 232 (2019).

The Web Below

1 Suzanne W. Simard and David A. Perry, "Net Transfer of Carbon between Ectomycorrhizal Tree Species in the Field," *Nature* 388, no. 6642 (1997).

Nutrients from Away

1 R.M. Peterman and B. Dorner, "Fraser River Sockeye Production Dynamics," Cohen Commission Technical Report (Vancouver: Cohen Commission of Inquiry into the Decline of Sockeye Salmon in the Fraser River, 2011).

2 Mark R. Gross, "Salmon Breeding Behavior and Life History Evolution in Changing Environments," *Ecology* 72, no. 4 (1991).

3 W.C. Clarke and T. Hirano, "Osmoregulation," in *Physiological Ecology of Pacific Salmon*, ed. L. Groot, L. Margolis, and W.C. Clarke (Vancouver: University of British Columba Press, 1995).

4 Dale Stokes, *The Fish in the Forest: Salmon and the Web of Life* (Berkeley: University of California Press, 2014).

Seabound

1 Ellen J. Censky, Karim Hodge, and Judy Dudley, "Over Water Dispersal of Lizards Due to Hurricanes," *Nature* 395, no. 6702 (1998).

2 Sherwin Carlquist, "Chance Dispersal: Long-Distance Dispersal of Organisms, Widely Accepted as Major Cause of Distribution Patterns, Poses Challenging Problems of Analysis," *American Scientist* 69, no. 5 (1981).

3 Homer, *The Odyssey*, trans. Robert Fagles (New York: Penguin Books, 1996), 167.

4 Chris T. Darimont, Paul C. Paquet, and Thomas E. Reimchen, "Landscape Heterogeneity and Marine Subsidy Generate Extensive Intrapopulation Niche Diversity in a Large Terrestrial Vertebrate," *The Journal of Animal Ecology* 78, no. 1 (January 2009).

5 For a compendium of island possibilities, see Rosemary Gillespie and David Clague, *Encyclopedia of Islands* (Berkeley: University of California Press, 2009).

The Route Finding of Lines

1 Alison Flood, "Oxford Junior Dictionary's Replacement of 'Natural' Words with 21st-Century Terms Sparks Outcry," *The Guardian*, January 13, 2015, https://www.

theguardian.com/books/2015/jan/13/oxford-junior-dictionary-replacement-natural-words.

2   John Berger, *Bento's Sketchbook* (New York: Pantheon Books, 2011), 150.

The Cost of Mobility

1   Anne Duputié and François Massol, "An Empiricist's Guide to Theoretical Predictions on the Evolution of Dispersal," *Interface Focus* 3, no. 6 (2013).

2   Andreas Schafer and David G. Victor, "The Future Mobility of the World Population," *Transportation Research Part A* 34 (2000).

3   Eric Zencey, "The Rootless Professors," in *Virgin Forest: Meditations on History, Ecology, and Culture* (Athens: University of Georgia Press, 1998), 61.

4   Reinhard F. Stettler, *Cottonwood and the River of Time: On Trees, Evolution and Society* (Seattle: University of Washington Press, 2009).

1   Michael Begon, Colin R. Townsend, and John L. Harper, *Ecology: From Individuals to Ecosystems*, 3rd ed. (Malden, MA: Blackwell Publishing, 1996), 469.

2   Henry A. Gleason, "Autobiographical Letter," *The Bulletin of the Ecological Society of America* 34, no. 2 (1953): 42.

3   For the origin of the Mongol horde metaphor, see Ragan M. Callaway and Wendy M. Ridenour, "Novel Weapons: Invasive Success and the Evolution of Increased Competitive Ability," *Frontiers in Ecology and the Environment* 2, no. 8 (2004).

4   Celestine Duncan, Jim Story, and Roger Sheley, *Biology, Ecology and Management of Montana Knapweeds* (Bozeman: Montana State University Extension, 2011), 2.

5   Compare previous report with Environmental Dynamics Inc., *Spotted Knapweed Species Account*, Forest Investment Account Land Based Investment Program, BC Ministry of Forests and Range, 2010, https://www.for.gov.bc.ca/hfd/library/FIA/2008/LBIP_2675015f.pdf.

6   Jim Jacobs, *Plant Guide for Spotted Knapweed (Centaurea Stoebe L.)* (Bozeman, MT: USDA-Natural Resources Conservation Service, 2012).

7   Paul J. Meiman, Edward F. Redente, and Mark W. Paschke, "The Role of the Native Soil Community in the Invasion Ecology of Spotted (*Centaurea maculosa* auct. non Lam.) and Diffuse (*Centaurea diffusa* Lam.) Knapweed," *Applied Soil Ecology* 32, no. 1 (May 2006).

Community Matters

1   Michael Begon, Colin R. Townsend, and John L. Harper, *Ecology: From Individuals to Ecosystems*, 3rd ed. (Malden, MA: Blackwell Publishing, 1996), 469.

2   Henry A. Gleason, "Autobiographical Letter," *The Bulletin of the Ecological Society of America* 34, no. 2 (1953): 42.

3   For the origin of the Mongol horde metaphor, see Ragan M. Callaway and Wendy M. Ridenour, "Novel Weapons: Invasive Success and the Evolution of Increased Competitive Ability," *Frontiers in Ecology and the Environment* 2, no. 8 (2004).

4 Celestine Duncan, Jim Story, and Roger Sheley, *Biology, Ecology and Management of Montana Knapweeds* (Bozeman: Montana State University Extension, 2011), 2.

5 Compare previous report with Environmental Dynamics Inc., *Spotted Knapweed Species Account*, Forest Investment Account Land Based Investment Program, BC Ministry of Forests and Range, 2010, https://www.for.gov.bc.ca/hfd/library/FIA/2008/LBIP_2675015f.pdf.

6 Jim Jacobs, *Plant Guide for Spotted Knapweed (Centaurea Stoebe L.)* (Bozeman, MT: USDA-Natural Resources Conservation Service, 2012).

7 Paul J. Meiman, Edward F. Redente, and Mark W. Paschke, "The Role of the Native Soil Community in the Invasion Ecology of Spotted (*Centaurea maculosa* auct. non Lam.) and Diffuse (*Centaurea diffusa* Lam.) Knapweed," *Applied Soil Ecology* 32, no. 1 (May 2006).

Mapping Moss

1 Andrea Wulf, *The Invention of Nature: Alexander von Humboldt's New Worlds* (New York: Alfred A. Knopf, 2015).

2 Jeanette Armstrong, *Breath Tracks* (Stratford, ON: Theytus Books, 1991), 28.

3 David Mackay, "Agents of Empire: The Banksian Collectors and Evaluations of New Lands," in *Visions of Empire: Voyages, Botany, and Representations of Nature*, ed. David Philip Miller and Peter Hanns Reill (Cambridge: Cambridge University Press, 1996), 38.

4 Suzanne Zeller, *Inventing Canada: Early Victorian Science and the Idea of a Transcontinental Nation* (Montreal: McGill-Queen's University Press, 2009).

5 Brittany Blachford, "Exploring the History of Women in Botany: Tracing Seven Female Contributors of the UBC Herbarium," University of British Columbia, GEOG 429, University of British Columbia Library, 2013. https://doi.org/10.14288/1.0075696.

Say the Names

1 Joseph Wood Krutch, *The Forgotten Peninsula: A Naturalist in Baja California* (Tucson: University of Arizona Press, 1986), 61.

2 Leo C. Cronquist and Arthur Hitchcock, *Flora of the Pacific Northwest: An Illustrated Manual* (Seattle: University of Washington Press, 1973), vii.

When Mountains Move

1 Michael Pollan, *The Botany of Desire: A Plant's Eye View of the World*, 1st ed. (New York: Random House, 2001).

2 Lewis and Maslin, *The Human Planet*.

3 Penelope Green, "Meet the Plantfluencers," *The New York Times*, November 8, 2018, https://www.nytimes.com/2018/11/08/style/08SILL.html.

4 Pyle, *The Thunder Tree*.

5 Mary Louise Pratt, *Imperial Eyes: Travel Writing and Transculturation* (New York: Routledge, 1992), 7.

6 Denis Cosgrove, *Mappings* (London, UK: Reaktion Books, 1999), 2.

7 For an introduction to this literature, see Daniela Bleichmar, *Visible Empire: Botanical Expeditions and Visual Culture in the Hispanic Enlightenment* (Chicago: University of Chicago Press, 2012); Victoria Dickenson, *Drawn from Life: Science and Art in the Portrayal of the New World* (Toronto: University of Toronto Press, 1998); Felix Driver and Luciana Martins, *Tropical Visions in an Age of Empire* (Chicago: University of Chicago Press, 2005); Barbara Maria Stafford, *Voyage into Substance: Art, Science, Nature, and the Illustrated Travel Account, 1760–1840* (Cambridge: MIT Press, 1984).

8 Trevor Goward and Cathie Hickson, *Nature Wells Gray* (Edmonton: Lone Pine Publishing, 1995).

A Pattern with Consequences

1 R.H. Waring and J.F. Franklin, "Evergreen Coniferous Forests of the Pacific Northwest," *Science* 204, no. 4400 (1979).

2 Campbell River Museum, "Logging in the Jungles," http://www.crmuseum.ca/logging-jungles.

3 For an introduction to southern BC weather, see Cliff A. Mass, *The Weather of the Pacific Northwest* (Seattle: University of Washington Press, 2008).

4 Waring and Franklin, "Evergreen Coniferous Forests of the Pacific Northwest."

5 W.J. Bond, "The Tortoise and the Hare: Ecology of Angiosperm Dominance and Gymnosperm Dominance," *Biological Journal of the Linnean Society* 36 (1989).

The Finish of Relationship

1 A. Knapp et al., "The Keystone Role of Bison in North American Tallgrass Prairie," *Bioscience* 49 (1999).

2 Ingela Alger et al., "Paternal Provisioning Results from Ecological Change," *Proceedings of the National Academy of Sciences* 117, no. 20 (2020).

3 Joanna Reid, "Grassland Debates: Conservation and Social Change in the Cariboo-Chilcotin, British Columbia" (PhD diss., University of British Columbia, 2010).

4 Victoria Anderson, *Creatures of Empire: How Domestic Animals Transformed Early America* (Oxford: Oxford University Press, 2004).

5 R.N. Mack and J.N. Thompson, "Evolution in Steppe with Few Large, Hooved Mammals," *American Naturalist* 119 (1982).

6 Dr. Vernon Brink (2007) quoted in Reid, "Grassland Debates," 8.

7 Don Gayton quoted in Michael Blackstock and Rhonda McAllister, "First Nations Perspectives on the Grasslands of the Interior of British Columbia," *Journal of Ecological Anthropology* 8, no. 1 (2004): 31.

The Collecting Basket

1 Håkan Rydin and John K. Jeglum, *The Biology of Peatlands, Second Edition*, Biology of Habitats Series (Oxford: Oxford University Press, 2013); W.H. MacKenzie and J.R. Moran, *Wetlands of British Columbia: A Guide to Identification* (Victoria: Research Branch, Ministry of Forests, 2004).
2 For a basic introduction to *Sphagnum* ecology, see Nico van Breemen, "How Sphagnum Bogs Down Other Plants," *Trends in Ecology & Evolution* 10, no. 7 (July 1995).
3 Nina H. Nielsen et al., "The Last Meal of Tollund Man: New Analyses of His Gut Content," *Antiquity* 95, no. 383 (2021).
4 David James Duncan, "River Teeth: An Introduction – River Teeth Journal," https://www.riverteethjournal.com/about-us/river-teeth-an-introduction.
5 For a good overview of BC paleoecology, see Richard Hebda, "British Columbia Vegetation and Climate History with Focus on 6 Ka BP," *Géographie Physique et Quaternaire* 49, no. 1 (1995).
6 Robert Macfarlane, *The Old Ways: A Journey on Foot* (London: Penguin, 2012), 193.

Wearing Red

1 Victoria Finlay, *Color: A Natural History of the Palette* (New York: Random House, 2002).
2 Marco Archetti et al., "Unravelling the Evolution of Autumn Colours: An Interdisciplinary Approach," *Trends in Ecology & Evolution* 24, no. 3 (2009).
3 David Webster Lee, *Nature's Palette: The Science of Plant Color* (Chicago: University of Chicago Press, 2007).

Winter's Bottleneck

1 Peter Marchand, *Life in the Cold*, 2nd ed. (Hanover, NH: University Press of New England, 1991); James Halfpenny and Roy Ozanne, *Winter: An Ecological Handbook* (Boulder, CO: Johnson Publishing Co., 1989).

Collecting the Grip

1 G. Thomas Tanselle, "A Rationale of Collecting," *Studies in Bibliography* 51 (1998).
2 Werner Muensterberger, *Collecting: An Unruly Passion: Psychological Perspectives* (Princeton: Princeton University Press, 2016), 2.
3 Yuval Noah Harari, *Sapiens: A Brief History of Mankind* (New York: Harper Collins, 2015), 43.
4 Lisbet Koerner, "Purposes of Linnean Travel: A Preliminary Research Report," in *Visions of Empire: Voyages, Botany, and Representations of Nature*, ed. David Philip Miller and Peter Hans Reill (Cambridge: Cambridge University Press, 1996), 125.

5 Roberta Parish, R. Coupe, Dennis Lloyd, and Joe Antos, *Plants of Southern Interior British Columbia* (Vancouver: Lone Pine Publishing, 1996), 207.

6 Barbara Thiers, *Herbarium: The Quest to Preserve and Classify the World's Plants* (Portland, OR: Timber Press, 2020).

7 Patricia L.M. Lang et al., "Using Herbaria to Study Global Environmental Change," *New Phytologist* 221, no. 1 (2019).

8 Edward O. Wilson, *Biophilia* (Cambridge: Harvard University Press, 1984).

9 E.B. White, *The Letters of E.B. White*, ed. M. White (New York: Harper Collins, 2006), 263–64.

10 See, for example, Leonard Bell, "Not Quite Darwin's Artist: The Travel Art of Augustus Earle," *Journal of Historical Geography* 43 (2014).

11 Matthew B. Crawford, *The World Beyond Your Head: On Becoming an Individual in an Age of Distraction*, reprint ed. (Toronto: Penguin, 2015).

12 Daniel P. Bebber et al., "Big Hitting Collectors Make Massive and Disproportionate Contribution to the Discovery of Plant Species," *Proceedings of the Royal Society B: Biological Sciences* 279, no. 1736 (2012).

Carrying Capacity

1 Pat Shipman, *The Animal Connection* (New York: W.W. Norton, 2011), 14–15.

2 Matthew E. Gompper, "The Dog-Human-Wildlife Interface: Assessing the Scope of the Problem," in *Free-Ranging Dogs and Wildlife Conservation*, ed. Matthew E. Gompper (Oxford: Oxford University Press, 2014), 11.

3 Guo-dong Wang et al., "The Genomics of Selection in Dogs and the Parallel Evolution between Dogs and Humans," *Nature Communications* 4, no. 1860 (2013), https://doi.org/10.1038/ncomms2814.

4 Brian Hare and Michael Tomasello, "Human-Like Social Skills in Dogs?" *Trends in Cognitive Science* 9, no. 9 (2005).

5 Shipman, *The Animal Connection*, 61.

6 Pat Shipman, *The Invaders: How Humans and Their Dogs Drove Neanderthals to Extinction* (Cambridge: Belknap Press, 2015), 226–33.

7 Gompper, "The Dog-Human-Wildlife Interface."

8 Miho Nagasaw et al., "Oxytocin-Gaze Positive Loop and the Coevolution of Human-Dog Bonds," *Science* 348, no. 6232 (2015).

9 Peter Korn, *Why We Make Things and Why It Matters* (Boston: Godine, 2013), 123.

10 John Berger, "Drawing Is Discovery," *New Statesman* 142, no. 5156 (2013): 53.

A Kind of Courage

1 Aldo Leopold, *A Sand County Almanac: With Other Essays on Conservation from Round River* (New York: Ballantine Books, 1970), 197.

2 Glenn Albrecht, "Solastalgia: A New Concept in Health and Identity," *PAN* 3 (2005).

3 Masashi Soga and Kevin J. Gaston, "Shifting Baseline Syndrome: Causes, Consequences, and Implications," *Frontiers in Ecology and the Environment* 16, no. 4 (2018).

4 Hannah Hinchman, *A Trail through Leaves: The Journal as a Path to Place* (New York: W.W. Norton and Company, 1997), 167.

5 James Hansen et al., "Earth's Energy Imbalance: Confirmation and Implications," *Science* 308, no. 5727 (2005); Gerald A. Meehl et al., "How Much More Global Warming and Sea Level Rise?" *Science* 307, no. 5716 (2005).

6 David Beerling, *The Emerald Planet: How Plants Changed Earth's History* (Oxford: Oxford University Press, 2007).

7 Stephen Jay Gould, *Wonderful Life: The Burgess Shale and the Nature of History* (New York: W.W. Norton, 1989), 48.

8 Vernon H. Heywood, "Plant Conservation in the Anthropocene – Challenges and Future Prospects," *Plant Diversity* 39 (2017): 319.

9 For a recent review of estimates, see Eimear Nic Lughadha et al., "Extinction Risk and Threats to Plants and Fungi," *Plants, People, Planet* 2, no. 5 (2020).

10 Quentin Cronk, "Plant Extinctions Take Time," *Science*, July 29, 2016.

11 Sarah P. Otto, "Adaptation, Speciation and Extinction in the Anthropocene," *Proceedings of the Royal Society B: Biological Sciences* 285, no. 1891 (2018).

Forest Refuge(e)

1 BC Wildfire Service, "Wildfire Averages," Province of British Columbia, https://www2. gov.bc.ca/gov/content/safety/wildfire-status/about-bcws/wildfire-statistics/wildfire-averages.

2 David R. Roberts and Andreas Hamann, "Glacial Refugia and Modern Genetic Diversity of 22 Western North American Tree Species," *Proceedings of the Royal Society B: Biological Sciences* 281, no. 1804 (2015).

3 Arthur Dyke, "Late Quaternary Vegetation History of Northern North America Based on Pollen, Macrofossil, and Faunal Remains," *Géographie Physique et Quaternaire* 59, no. 2–3 (2005).

4 Anthony Ricciardi and Daniel Simberloff, "Assisted Colonization Is Not a Viable Conservation Strategy," *Trends in Ecology & Evolution* 24, no. 5 (2009).

5 Sally N. Aitken and Michael C. Whitlock, "Assisted Gene Flow to Facilitate Local Adaptation to Climate Change," *Annual Review of Ecology, Evolution, and Systematics* 44, no. 1 (2013).

6 Jason S. Mclachlan, Jessica J. Hellmann, and Mark W. Schwartz, "A Framework for Debate of Assisted Migration in an Era of Climate Change," *Conservation Biology* 21, no. 2 (2007): 299.

7 Oliver Kelhammer, Neo Eocene (Project) | Oliverk:Projects, http://www.oliverk.org/art-projects/land-art/neo-eocene-project.

8 Nicole L. Klenk and Brendon M.H. Larson, "The Assisted Migration of Western Larch in British Columbia: A Signal of Institutional Change in Forestry in Canada?" *Global Environmental Change* 31 (2015).

9 Crowther et al., "Mapping Tree Density at a Global Scale."

10 Intergovernmental Panel on Climate Change, "An IPCC Special Report on the Impacts of Global Warming of 1.5°C above Pre-Industrial Levels and Related Global Greenhouse Gas Emission Pathways," 2018, https://www.ipcc.ch/sr15/.

11 Jean-Francois Bastin et al., "The Global Tree Restoration Potential," *Science* 365, no. 6448 (2019).

An Unquiet Botany

1 For the first use of this term, see Christian Rutz et al., "COVID-19 Lockdown Allows Researchers to Quantify the Effects of Human Activity on Wildlife," *Nature Ecology & Evolution* 4, no. 9 (2020).

Straddling Worlds

1 Trevor Goward, *North by Northwest*, interview by Sheryl MacKay, CBC Radio, June 13, 2021, https://www.cbc.ca/listen/live-radio/1-43-north-by-northwest/clip/15851374-sunday-june-13.

2 Toby Spribille et al., "Basidiomycete Yeasts in the Cortex of Ascomycete Macrolichens," *Science (New York, NY)* 353, no. 6298 (2016).

3 Hina Alam, "80% of Mountain Glaciers in Alberta, BC and Yukon Will Disappear within 50 Years," *CBC News*, December 27, 2018, https://www.cbc.ca/news/canada/british-columbia/western-glaciers-disappear-50-years-1.4959663.

4 Don McKay, "The Incarnation of Betweenity: Contemplating Lichens," presented at Enlivenment, Enlichenment and the Poetics of Place, Wells Gray Education and Research Centre, Upper Clearwater Valley, April 24, 2017.

5 Trevor Goward, "Ways of Enlichenment – Twelve Readings on the Lichen Thallus," 2011, https://www.waysofenlichenment.net/ways/readings/index.

6 Jeanette Armstrong, *Indigenization*, video, TEDxOkanagan College, 2011, https://www.youtube.com/watch?v=jLOfXsFlb18.

7 Martin Lee Mueller, *Being Salmon, Being Human: Encountering the Wild in Us and Us in the Wild* (White River Junction, VT: Chelsea Green, 2017), 191.

8 Robert Bringhurst, *The Tree of Meaning: Language, Mind and Ecology* (Berkeley, CA: Counterpoint Press, 2008), 275.

9 Aristotle, *The Nicomachean Ethics*, ed. Lesley Brown (Oxford: Oxford University Press, 2009), 1103b1, 92.

1 See, for example, Trefor B. Reynoldson et al., "Fraser River Basin," in *Rivers of North America*, ed. Arthur C. Benke and Colbert E. Cushing (Amsterdam, Netherlands: Elsevier Academic Press, 2005).

2 Alexander Koch et al., "Earth System Impacts of the European Arrival and Great Dying in the Americas after 1492," *Quaternary Science Reviews* 207 (2019).

3 Marianne Ignace and Ronald E. Ignace, *Secwépemc People, Land, and Laws* (Montreal: McGill-Queen's University Press, 2017).

4 Ronald E. Ignace and Marianne B. Ignace, "Re Tsuwet.s-Kucw Ne Secwepemculecw: Secwepemc Resource Use and Sense of Place," in *Secwepemc People and Plants: Research Papers in Shuswap Ethnobotany*, ed. Marianne B. Ignace, Nancy J. Turner, and Sandra J. Peacock (Seattle: Society of Ethnobiology, 2016), https://ethnobiology.org/publications/contributions/secwepemc-people-and-plants-research-papers-shuswap-ethnobotany.

5 D'Arcy Jenish, *Epic Wanderer: David Thompson and the Mapping of the Canadian West* (Toronto: Doubleday Canada, 2003).

6 Caitlin Quist, *Public Produce and the Butler Urban Farm* (Kamloops: Kamloops Food Policy Council, 2021), https://kamloopsfoodpolicycouncil.com/wp-content/uploads/2021/03/PublicProduceTheButlerUrbanFarm.pdf.

7 National Inquiry into Missing and Murdered Indigenous Women and Girls, *Reclaiming Power and Place: The Final Report of the National Inquiry into Missing and Murdered Indigenous Women and Girls*, vol. 1a, 2019, https://www.mmiwg-ffada.ca/wp-content/uploads/2019/06/Final_Report_Vol_1a-1.pdf; First Nation Information Governance Centre, *National Report of the First Nations Regional Health Survey*, 2018, https://fnigc.ca/wp-content/uploads/2020/09/fnigc_rhs_phase_3_volume_two_en_final_screen.pdf; Human Rights Watch, "My Fear Is Losing Everything": The Climate Crisis and First Nations' Right to Food in Canada, 2020, https://www.hrw.org/report/2020/10/21/my-fear-losing-everything/climate-crisis-and-first-nations-right-food-canada.

8 See the description of the 1862–1863 smallpox epidemic in Ignace and Ignace, *Secwépemc People, Land, and Laws*.

9 Ken Farvoldt, "Learning about the Legendary Overlanders," *Kamloops This Week*, July 29, 2020, https://www.kamloopsthisweek.com/community/history-learning-about-the-legendary-overlanders-4445447.

10 See Laura Lawson, *City Bountiful: A Century of Community Gardening in America* (Berkeley: University of California Press, 2005); Diana Mincyte and Karin Dobernig, "Urban Farming in the North American Metropolis: Rethinking Work and Distance in Alternative Food Networks," *Environment and Planning A: Economy and Space* 48, no. 9 (2016); Bonnie J. Clark, *Finding Solace in the Soil: An Archaeology of Gardens and Gardeners at Amache*, 1st ed. (Louisville: University Press of Colorado, 2020); Kelsey Timler and Dancing Water Sandy, "Gardening in Ashes: The Possibilities and Limitations of Gardening to Support Indigenous Health and Well-Being in the Context of Wildfires and Colonialism," *International Journal of Environmental Research and Public Health* 17,

no. 9 (2020); Oliver Sacks, "The Healing Power of Gardens," *The New York Times*, April 18, 2019, https://www.nytimes.com/2019/04/18/opinion/sunday/oliver-sacks-gardens.html.

Reconciling Botany Pond

1   Arthur Manuel, Naomi Klein, and R.M. Derrickson, *Unsettling Canada: A National Wake-up Call* (Toronto: Between the Lines, 2015), 1–10.
2   J. Edward Chamberlin, *If This Is Your Land, Where Are Your Stories?* (Cleveland, OH: Pilgrim Press, 2003), 78.
3   Lili Anolik, "Money, Madness, Cocaine and Literary Genius: An Oral History of the 1980s' Most Decadent College," *Esquire*, May 28, 2019, https://www.esquire.com/entertainment/a27434009/bennington-college-oral-history-bret-easton-ellis/.
4   John Keats, "Ode to a Nightingale," in *The Norton Anthology of Poetry, Revised Edition*, ed. Alexander W. Allison, Herbert Barrows, Caesar R. Blake, Arthur J. Carr, Arthur M. Eastman, and Hubert M. English Jr. (New York: W.W. Norton and Company, 1975), 711.
5   Richard Hugo, *The Lady in Kicking Horse Reservoir: Poems* (New York: W.W. Norton and Company, 1973) 9, 61, 73.
6   Wes Jackson, *Becoming Native to This Land* (Berkeley, CA: Counterpoint Press, 1996), 3.
7   Cole Harris, "How Did Colonialism Dispossess? Comments from an Edge of Empire," *Annals of the Association of American Geographers* 94, no. 1 (2004): 174.
8   Murray Sinclair, "Truth and Reconciliation," presented at the Presidential Lecture Series/7th Storytellers Gala, Thompson Rivers University, Kamloops, BC, October 26, 2015.
9   Ignace and Ignace, *Secwépemc People, Land, and Laws*, 46.
10 Manuel, Klein, and Derrickson, *Unsettling Canada: A National Wake-up Call.*
11 Roland Neaves, "A History of Settlement in the Lac Du Bois Basin, 1840–1971: A Study in Sequent Occupance," *BC Perspectives* 1 (1972).
12 Donna Haraway, *Staying with the Trouble: Making Kin in the Chthulucene* (Durham, NC: Duke University Press, 2016).

Form Follows Function

1   Richard Wagamese, *One Native Life* (Madeira Park, BC: Douglas and MacIntyre, 2008), 4.
2   Thomas King, *The Truth about Stories* (Toronto: House of Anansi Press, 2003), 164.

Aitken, Sally N., and Michael C. Whitlock. "Assisted Gene Flow to Facilitate Local Adaptation to Climate Change." *Annual Review of Ecology, Evolution, and Systematics* 44, no. 1 (2013): 367–88.

Alam, Hina. "80% of Mountain Glaciers in Alberta, BC and Yukon Will Disappear within 50 Years." *CBC News*, December 27, 2018. https://www.cbc.ca/news/canada/british-columbia/western-glaciers-disappear-50-years-1.4959663.

Albrecht, Glenn. "Solastalgia: A New Concept in Health and Identity." *PAN* 3 (2005): 41–55.

Alger, Ingela, Paul L. Hooper, Donald Cox, Jonathan Stieglitz, and Hillard S. Kaplan. "Paternal Provisioning Results from Ecological Change." *Proceedings of the National Academy of Sciences* 117, no. 20 (2020): 10746–54.

Anderson, Victoria. *Creatures of Empire: How Domestic Animals Transformed Early America.* Oxford: Oxford University Press, 2004.

Anolik, Lili. "Money, Madness, Cocaine and Literary Genius: An Oral History of the 1980s' Most Decadent College." *Esquire*, May 28, 2019. https://www.esquire.com/entertainment/a27434009/bennington-college-oral-history-bret-easton-ellis/.

Archetti, Marco, Thomas F. Döring, Snorre B. Hagen, Nicole M. Hughes, Simon R. Leather, David W. Lee, Simcha Lev-Yadun, et al. "Unravelling the Evolution of Autumn Colours: An Interdisciplinary Approach." *Trends in Ecology & Evolution* 24, no. 3 (2009): 166–73.

Aristotle. *The Nicomachean Ethics.* Edited by Lesley Brown. Oxford: Oxford University Press, 2009.

Armstrong, Jeannette. *Breath Tracks.* Stratford, ON: Theytus Books, 1991.

Armstrong, Jeannette. *Indigenization.* Video. TEDxOkanagan College, 2011. https://www.youtube.com/watch?v=jLOfXsFlb18.

Baldwin, Lyn K. "Drawing Care: The Illustrated Journal's 'Path to Place.'" *Journal of Teaching in Travel & Tourism* 18, no. 1 (2018): 75–93.

Bar-On, Yinon M., Rob Phillips, and Ron Milo. "The Biomass Distribution on Earth." *Proceedings of the National Academy of Sciences* 115, no. 25 (2018): 6506–11.

Bastin, Jean-Francois, Yelena Finegold, Claude Garcia, Danilo Mollicone, Marcelo Rezende, Devin Routh, Constantin M. Zohner, et al. "The Global Tree Restoration Potential." *Science* 365, no. 6448 (2019): 76–79.

BC Wildfire Service. "Wildfire Averages." Province of British Columbia. https://www2.gov.bc.ca/gov/content/safety/wildfire-status/about-bcws/wildfire-statistics/wildfire-averages.

Beach, Sarah. "Curse of the Hippie Parents." *Salon*, August 22, 2001. https://www.salon.com/2001/08/22/hippie_parents/.

Bebber, Daniel P., Mark A. Carine, Gerrit Davidse, David J. Harris, Elspeth M. Haston, Malcolm G. Penn, Steve Cafferty, John R.I. Wood, and Robert W. Scotland. "Big Hitting Collectors Make Massive and Disproportionate Contribution to the Discovery of Plant Species." *Proceedings of the Royal Society B: Biological Sciences* 279, no. 1736 (2012): 2269–74.

Beerling, David. *The Emerald Planet: How Plants Changed Earth's History.* Oxford: Oxford University Press, 2007.

Begon, Michael, Colin R. Townsend, and John L. Harper. *Ecology: From Individuals to Ecosystems.* 3rd Edition. Malden, MA: Blackwell Publishing, 1996.

Bell, Leonard. "Not Quite Darwin's Artist: The Travel Art of Augustus Earle." *Journal of Historical Geography* 43 (2014): 60–70.

Berger, John. *Bento's Sketchbook.* New York: Pantheon Books, 2011.

Berger, John. "Drawing Is Discovery." *New Statesman* 142, no. 5156 (2013): 53.

Blachford, Brittany. "Exploring the History of Women in Botany: Tracing Seven Female Contributors of the UBC Herbarium." University of British Columbia. GEOG 429. University of British Columbia Library, 2013. https://doi.org/10.14288/1.0075696.

Blackstock, Michael, and Rhonda McAllister. "First Nations Perspectives on the Grasslands of the Interior of British Columbia." *Journal of Ecological Anthropology* 8, no. 1 (2004): 24–46.

Bleichmar, Daniela. *Visible Empire: Botanical Expeditions and Visual Culture in the Hispanic Enlightenment.* Chicago: University of Chicago Press, 2012.

Bond, W.J. "The Tortoise and the Hare: Ecology of Angiosperm Dominance and Gymnosperm Dominance." *Biological Journal of the Linnean Society* 36 (1989): 227–49.

Breemen, Nico van. "How Sphagnum Bogs Down Other Plants." *Trends in Ecology & Evolution* 10, no. 7 (July 1995): 270–75.

Bringhurst, Robert. *The Tree of Meaning: Language, Mind and Ecology.* Berkeley, CA: Counterpoint Press, 2008.

Brown, Justine. *All Possible Worlds: Utopian Experiments in British Columbia.* Vancouver: New Star Books, 1995.

Callaway, Ragan M., and Wendy M. Ridenour. "Novel Weapons: Invasive Success and the Evolution of Increased Competitive Ability." *Frontiers in Ecology and the Environment* 2, no. 8 (2004): 436–43.

Campbell River Museum. "Logging in the Jungles." https://crmuseum.ca/2019/11/28/logging-in-the-jungles/.

Carlquist, Sherwin. "Chance Dispersal: Long-Distance Dispersal of Organisms, Widely Accepted as Major Cause of Distribution Patterns, Poses Challenging Problems of Analysis." *American Scientist* 69, no. 5 (1981): 509–16.

Censky, Ellen J., Karim Hodge, and Judy Dudley. "Over-Water Dispersal of Lizards Due to Hurricanes." *Nature* 395, no. 6702 (1998): 556.

Chamberlin, J. Edward. *If This Is Your Land, Where Are Your Stories?* Cleveland, OH: Pilgrim Press, 2003.

Clark, Bonnie J. *Finding Solace in the Soil: An Archaeology of Gardens and Gardeners at Amache.* 1st Edition. Louisville: University Press of Colorado, 2020.

Clarke, W.C., and T. Hirano. "Osmoregulation." In *Physiological Ecology of Pacific Salmon*, edited by L. Groot, L. Margolis, and W.C. Clarke, 317–77. Vancouver: University of British Columba Press, 1995.

Cosgrove, Denis. *Mappings.* London, UK: Reaktion Books, 1999.

Crawford, Matthew B. *The World Beyond Your Head: On Becoming an Individual in an Age of Distraction.* Reprint Edition. Toronto: Penguin, 2015.

Cronk, Quentin. "Plant Extinctions Take Time." *Science*, July 29, 2016, 446–47.

Cronquist, C. Leo, and Arthur Hitchcock. *Flora of the Pacific Northwest: An Illustrated Manual.* Seattle: University of Washington Press, 1973.

Crowther, T.W., H.B. Glick, K.R. Covey, C. Bettigole, D.S. Maynard, S.M. Thomas, J.R. Smith, et al. "Mapping Tree Density at a Global Scale." *Nature* 525, no. 7568 (2015): 201–5.

Darimont, Chris T., Paul C. Paquet, and Thomas E. Reimchen. "Landscape Heterogeneity and Marine Subsidy Generate Extensive Intrapopulation Niche Diversity in a Large Terrestrial Vertebrate." *The Journal of Animal Ecology* 78, no. 1 (January 2009): 126–33.

Dickenson, Victoria. *Drawn from Life: Science and Art in the Portrayal of the New World.* Toronto: University of Toronto Press, 1998.

Driver, Felix, and Luciana Martins. *Tropical Visions in an Age of Empire.* Chicago: University of Chicago Press, 2005.

Duncan, Celestine, Jim Story, and Roger Sheley. *Biology, Ecology and Management of Montana Knapweeds.* Bozeman: Montana State University Extension, 2011.

Duncan, David James. "River Teeth: An Introduction – River Teeth Journal." https://www.riverteethjournal.com/about-us/river-teeth-an-introduction.

Duputié, Anne, and François Massol. "An Empiricist's Guide to Theoretical Predictions on the Evolution of Dispersal." *Interface Focus* 3, no. 6 (2013): 20130028.

Dyke, Arthur. "Late Quaternary Vegetation History of Northern North America Based on Pollen, Macrofossil, and Faunal Remains." *Géographie Physique et Quaternaire* 59, no. 2–3 (2005): 211–62.

Ellison, Aaron M., Carri J. LeRoy, Kim J. Landsbergen, Emily Bosanquet, David Buckley Borden, Paul J. CaraDonna, Katherine Cheney, et al. "Art/Science Collaborations: New Explorations of Ecological Systems, Values, and Their Feedbacks." *The Bulletin of the Ecological Society of America* 99, no. 2 (2018): 180–91.

Environmental Inc. *Spotted Knapweed Species Account.* Forest Investment Account Land Based Investment Program. BC Ministry of Forests and Range, 2010. https://www.for.gov.bc.ca/hfd/library/FIA/2008/LBIP_2675015f.pdf.

Farvoldt, Ken. "Learning about the Legendary Overlanders." *Kamloops This Week*, July 29, 2020. https://www.kamloopsthisweek.com/community/history-learning-about-the-legendary-overlanders-4445447.

Finlay, Victoria. *Color: A Natural History of the Palette.* New York: Random House, 2002.

First Nation Information Governance Centre. *National Report of the First Nations Regional Health Survey.* 2018. https://fnigc.ca/wp-content/uploads/2020/09/fnigc_rhs_phase_3_volume_two_en_final_screen.pdf.

Flood, Alison. "Oxford Junior Dictionary's Replacement of 'Natural' Words with 21st-Century Terms Sparks Outcry." *The Guardian*, January 13, 2015. https://www.theguardian.com/books/2015/jan/13/oxford-junior-dictionary-replacement-natural-words.

Gillespie, Rosemary, and David Clague. *Encyclopedia of Islands.* Berkeley: University of California Press, 2009.

Gleason, Henry A. "Autobiographical Letter." *The Bulletin of the Ecological Society of America* 34, no. 2 (1953): 40–42.

Gompper, Matthew E. "The Dog-Human-Wildlife Interface: Assessing the Scope of the Problem." In *Free-Ranging Dogs and Wildlife Conservation*, edited by Matthew E. Gompper, 15–25. Oxford: Oxford University Press, 2014.

Gould, Rebecca Kneale. "Back to the Land Movements." In *Encyclopedia of Religion and Nature*, edited by Bron Taylor, 149–51. London: Continuum, 2005.

Gould, Stephen Jay. *Wonderful Life: The Burgess Shale and the Nature of History.* New York: W.W. Norton, 1989.

Goward, Trevor. *North by Northwest.* Interview by Sheryl MacKay. CBC Radio, June 13, 2021. https://www.cbc.ca/listen/live-radio/1-43-north-by-northwest/clip/15851374-sunday-june-13.

Goward, Trevor. "Ways of Enlichenment – Twelve Readings on the Lichen Thallus." 2011. https://www.waysofenlichenment.net/ways/readings/index.

Goward, Trevor, and Cathie Hickson. *Nature Wells Gray.* Edmonton: Lone Pine Publishing, 1995.

Green, Penelope. "Meet the Plantfluencers." *The New York Times*, November 8, 2018, https://www.nytimes.com/2018/11/08/style/08SILL.html.

Gross, Mark R. "Salmon Breeding Behavior and Life History Evolution in Changing Environments. *Ecology* 72, no. 4 (1991): 1180–86.

Halfpenny, James, and Roy Ozanne. *Winter: An Ecological Handbook.* Boulder, CO: Johnson Publishing Co., 1989.

Hansen, James, Larissa Nazarenko, Reto Ruedy, Makiko Sato, Josh Willis, Anthony Del Genio, Dorothy Koch, et al. "Earth's Energy Imbalance: Confirmation and Implications." *Science* 308, no. 5727 (2005): 1431–35.

Harari, Yuval Noah. *Sapiens: A Brief History of Mankind.* New York: Harper Collins, 2015.

Haraway, Donna. *Staying with the Trouble: Making Kin in the Chthulucene.* Durham, NC: Duke University Press, 2016.

Hare, Brian, and Michael Tomasello. "Human-Like Social Skills in Dogs?" *Trends in Cognitive Science* 9, no. 9 (2005): 439–44.

Harris, Cole. "How Did Colonialism Dispossess? Comments from an Edge of Empire." *Annals of the Association of American Geographers* 94, no. 1 (2004): 165–82.

Hebda, Richard. "British Columbia Vegetation and Climate History with Focus on 6 Ka BP." *Géographie Physique et Quaternaire* 49, no. 1 (1995): 55–79.

Heywood, Vernon, H. "Plant Conservation in the Anthropocene – Challenges and Future Prospects." *Plant Diversity* 39 (2017): 314–30.

Hinchman, Hannah. *A Trail through Leaves: The Journal as a Path to Place.* New York: W.W. Norton and Company, 1997.

Homer. *The Odyssey.* Translated by Robert Fagles. New York: Penguin Books, 1996.

Hugo, Richard. *The Lady in Kicking Horse Reservoir: Poems.* New York: W.W. Norton and Company, 1973.

Human Rights Watch. *"My Fear Is Losing Everything": The Climate Crisis and First Nations' Right to Food in Canada.* 2020. https://www.hrw.org/report/2020/10/21/my-fear-losing-everything/climate-crisis-and-first-nations-right-food-canada.

Ignace, Marianne, and Ronald E. Ignace. *Secwépemc People, Land, and Laws.* Montreal: McGill-Queen's University Press, 2017.

Ignace, Ronald E., and Marianne B. Ignace. "Re Tsuwet.s-Kucw Ne Secwepemculecw: Secwepemc Resource Use and Sense of Place." In *Secwepemc People and Plants: Research Papers in Shuswap Ethnobotany*, edited by Marianne B. Ignace, Nancy J. Turner, and Sandra J. Peacock, 25–62. Seattle: Society of Ethnobiology, 2016. https://ethnobiology.org/publications/contributions/secwepemc-people-and-plants-research-papers-shuswap-ethnobotany.

Intergovernmental Panel on Climate Change. "An IPCC Special Report on the Impacts of Global Warming of 1.5°C above Pre-Industrial Levels and Related Global Greenhouse Gas Emission Pathways." 2018. https://www.ipcc.ch/sr15/.

Jackson, Wes. *Becoming Native to This Land.* Berkeley, CA: Counterpoint Press, 1996.

Jacobs, Jim. *Plant Guide for Spotted Knapweed (Centaurea Stoebe L.).* Bozeman, MT: USDA-Natural Resources Conservation Service, 2012.

Jenish, D'Arcy. *Epic Wanderer: David Thompson and the Mapping of the Canadian West.* Toronto: Doubleday Canada, 2003.

Keats, John. "Ode to a Nightingale." In *The Norton Anthology of Poetry, Revised Edition*, edited by Alexander W. Allison, Herbert Barrows, Caesar R. Blake, Arthur J. Carr, Arthur M. Eastman, and Hubert M. English Jr., 711. New York: W.W. Norton and Company, 1975.

Kelhammer, Oliver. Neo Eocene (Project) | Oliverk:Projects. http://www.oliverk.org/art-projects/land-art/neo-eocene-project.

King, Thomas. *The Truth about Stories.* Toronto: House of Anansi Press, 2003.

Klenk, Nicole L., and Brendon M.H. Larson. "The Assisted Migration of Western Larch in British Columbia: A Signal of Institutional Change in Forestry in Canada?" *Global Environmental Change* 31 (2015): 20–27.

Knapp, A., J. Blair, S. Collins, D. Hartnett, L. Johnson, and E.G. Town. "The Keystone Role of Bison in North American Tallgrass Prairie." *Bioscience* 49 (1999): 40–50.

Koch, Alexander, Chris Brierley, Mark M. Maslin, and Simon L. Lewis. "Earth System Impacts of the European Arrival and Great Dying in the Americas after 1492." *Quaternary Science Reviews* 207 (2019): 13–36.

Koerner, Lisbet. "Purposes of Linnean Travel: A Preliminary Research Report." In *Visions of Empire: Voyages, Botany, and Representations of Nature*, edited by David Philip Miller and Peter Hans Reill, 117–52. Cambridge: Cambridge University Press, 1996.

Korn, Peter. *Why We Make Things and Why It Matters*. Boston: Godine, 2013.

Krutch, Joseph Wood. *The Forgotten Peninsula: A Naturalist in Baja California*. Tucson: University of Arizona Press, 1986.

Lang, Patricia L.M., Franziska M. Willems, J.F. Scheepens, Hernán A. Burbano, and Oliver Bossdorf. "Using Herbaria to Study Global Environmental Change." *New Phytologist* 221, no. 1 (2019): 110–22.

Lawson, Laura. *City Bountiful: A Century of Community Gardening in America*. Berkeley: University of California Press, 2005.

Lee, David Webster. *Nature's Palette: The Science of Plant Color*. Chicago: University of Chicago Press, 2007.

Leopold, Aldo. *A Sand County Almanac: With Other Essays on Conservation from Round River*. New York: Ballantine Books, 1970.

Lewis, Simon L., and Mark A. Maslin. *The Human Planet: How We Created the Anthropocene*. London: Pelican Books, 2018.

Macfarlane, Robert. *The Old Ways: A Journey on Foot*. London: Penguin, 2012.

Mack, R.N., and J.N. Thompson. "Evolution in Steppe with Few Large, Hooved Mammals." *American Naturalist* 119 (1982): 757–73.

Mackay, David. "Agents of Empire: The Banksian Collectors and Evaluations of New Lands." In *Visions of Empire: Voyages, Botany, and Representations of Nature*, edited by David Philip Miller and Peter Hanns Reill, 38–57. Cambridge: Cambridge University Press, 1996.

MacKenzie, W.H., and J.R. Moran. *Wetlands of British Columbia: A Guide to Identification*. Victoria: Research Branch, Ministry of Forests, 2004.

Manuel, Arthur, Naomi Klein, and R.M. Derrickson. *Unsettling Canada: A National Wake-up Call*. Toronto: Between the Lines, 2015.

Marchand, Peter. *Life in the Cold*. 2nd Edition. Hanover, NH: University Press of New England, 1991.

Mass, Cliff A. *The Weather of the Pacific Northwest*. Seattle: University of Washington Press, 2008.

McKay, Don. "The Incarnation of Betweenity: Contemplating Lichens." Presented at Enlivenment, Enlichenment and the Poetics of Place, Wells Gray Education and Research Centre, Upper Clearwater Valley, April 24, 2017.

Mclachlan, Jason S., Jessica J. Hellmann, and Mark W. Schwartz. "A Framework for Debate of Assisted Migration in an Era of Climate Change." *Conservation Biology* 21, no. 2 (2007): 297–302.

Meehl, Gerald A., Warren M. Washington, William D. Collins, Julie M. Arblaster, Aixue Hu, Lawrence E. Buja, Warren G. Strand, et al. "How Much More Global Warming and Sea Level Rise?" *Science* 307, no. 5716 (2005): 1769–72.

Meiman, Paul J., Edward F. Redente, and Mark W. Paschke. "The Role of the Native Soil Community in the Invasion Ecology of Spotted (*Centaurea maculosa* auct. non Lam.) and Diffuse (*Centaurea diffusa* Lam.) Knapweed." *Applied Soil Ecology* 32, no. 1 (May 2006): 77–88.

Mincyte, Diana, and Karin Dobernig. "Urban Farming in the North American Metropolis: Rethinking Work and Distance in Alternative Food Networks." *Environment and Planning A: Economy and Space* 48, no. 9 (2016): 1767–86.

Mueller, Martin Lee. *Being Salmon, Being Human: Encountering the Wild in Us and Us in the Wild.* White River Junction, VT: Chelsea Green, 2017.

Muensterberger, Werner. *Collecting: An Unruly Passion: Psychological Perspectives.* Princeton: Princeton University Press, 2016.

Myers, Natasha. "From Edenic Apocalypse to Gardens against Eden: Plants and People in and after the Anthropocene." In *Infrastructure, Environment, and Life in the Anthropocene*, edited by Kregg Hetherington, 115–48. Durham, NC: Duke University Press, 2019.

Nagasaw, Miho, Shouhei Mitsui, Shiori En, Nobuyo Ohtani, Mitsuaki Ohta, Yasuo Sakuma, Tatsushi Onaka, et al. "Oxytocin-Gaze Positive Loop and the Coevolution of Human-Dog Bonds." *Science* 348, no. 6232 (2015): 333–36.

National Inquiry into Missing and Murdered Indigenous Women and Girls. *Reclaiming Power and Place: The Final Report of the National Inquiry into Missing and Murdered Indigenous Women and Girls.* Volume 1a, 2019. https://www.mmiwg-ffada.ca/wp-content/uploads/2019/06/Final_Report_Vol_1a-1.pdf.

Neaves, Roland. "A History of Settlement in the Lac Du Bois Basin, 1840–1971: A Study in Sequent Occupance." *BC Perspectives* 1 (1972): 4–23.

Nic Lughadha, Eimear, Steven P. Bachman, Tarciso C.C. Leão, Félix Forest, John M. Halley, Justin Moat, Carmen Acedo, et al. "Extinction Risk and Threats to Plants and Fungi." *Plants, People, Planet* 2, no. 5 (2020): 389–408.

Nielsen, Nina H., Peter Steen Henriksen, Morten Fischer Mortensen, Renée Enevold, Martin N. Mortensen, Carsten Scavenius, and Jan J. Enghild. "The Last Meal of Tollund Man: New Analyses of His Gut Content." *Antiquity* 95, no. 383 (2021): 1195–1212.

Otto, Sarah P. "Adaptation, Speciation and Extinction in the Anthropocene." *Proceedings of the Royal Society B: Biological Sciences* 285, no. 1891 (2018): 20182047.

Parish, Roberta, R. Coupe, Dennis Lloyd, and Joe Antos. *Plants of Southern Interior British Columbia.* Vancouver: Lone Pine Publishing, 1996.

Peterman, R.M., and B. Dorner. "Fraser River Sockeye Production Dynamics." Cohen Commission Technical Report. Vancouver: Cohen Commission of Inquiry into the Decline of Sockeye Salmon in the Fraser River, 2011.

Pollan, Michael. *The Botany of Desire: A Plant's Eye View of the World.* 1st Edition. New York: Random House, 2001.

Pratt, Mary Louise. *Imperial Eyes: Travel Writing and Transculturation.* New York: Routledge, 1992.

Pyle, Robert Michael. *The Thunder Tree: Lessons from an Urban Wildland.* New York: Lyons Press, 1993.

Quist, Caitlin. *Public Produce and the Butler Urban Farm.* Kamloops: Kamloops Food Policy Council, 2021. https://kamloopsfoodpolicycouncil.com/wp-content/uploads/2021/03/PublicProduceTheButlerUrbanFarm.pdf.

Reid, Joanna. "Grassland Debates: Conservation and Social Change in the Cariboo-Chilcotin, British Columbia." PhD diss., University of British Columbia, 2010.

Reynoldson, Trefor B., Joseph Culp, Rick Lowell, and John C. Richardson. "Fraser River Basin." In *Rivers of North America*, edited by Arthur C. Benke and Colbert E. Cushing, 697–732. Amsterdam, Netherlands: Elsevier Academic Press, 2005.

Ricciardi, Anthony, and Daniel Simberloff. "Assisted Colonization Is Not a Viable Conservation Strategy." *Trends in Ecology & Evolution* 24, no. 5 (2009): 248–53.

Roberts, David R., and Andreas Hamann. "Glacial Refugia and Modern Genetic Diversity of 22 Western North American Tree Species." *Proceedings of the Royal Society B: Biological Sciences* 282, no. 1804 (2015): 20142903.

Rutz, Christian, Matthias-Claudio Loretto, Amanda E. Bates, Sarah C. Davidson, Carlos M. Duarte, Walter Jetz, Mark Johnson, et al. "COVID-19 Lockdown Allows Researchers to Quantify the Effects of Human Activity on Wildlife." *Nature Ecology & Evolution* 4, no. 9 (2020): 1156–59.

Rydin, Håkan, and John K. Jeglum. *The Biology of Peatlands, Second Edition*. Biology of Habitats Series. Oxford: Oxford University Press, 2013.

Sacks, Oliver. "The Healing Power of Gardens." *The New York Times*, April 18, 2019. https://www.nytimes.com/2019/04/18/opinion/sunday/oliver-sacks-gardens.html.

Sánchez-Bayo, Francisco, and Kris A.G. Wyckhuys. "Worldwide Decline of the Entomofauna: A Review of Its Drivers." *Biological Conservation* 232 (2019): 8–27.

Schafer, Andreas, and David G. Victor. "The Future Mobility of the World Population." *Transportation Research Part A* 34 (2000): 171–205.

Schauffler, F. Marina. *Turning to Earth: Stories of Ecological Conversion*. Charlottesville: University of Virginia Press, 2003.

Shipman, Pat. *The Animal Connection*. New York: W.W. Norton, 2011.

Shipman, Pat. *The Invaders: How Humans and Their Dogs Drove Neanderthals to Extinction*. Cambridge: Belknap Press, 2015.

Simard, Suzanne W., and David A. Perry. "Net Transfer of Carbon between Ectomycorrhizal Tree Species in the Field." *Nature* 388, no. 6642 (1997): 579.

Simmons, Terry Allen. "But We Must Cultivate Our Garden: Twentieth Century Pioneering in Rural British Columbia." PhD diss., University of Minnesota, 1979.

Sinclair, Murray. "Truth and Reconciliation." Presented at the Presidential Lecture Series/7th Storytellers Gala, Thompson Rivers University, Kamloops, BC, October 26, 2015.

Soga, Masashi, and Kevin J. Gaston. "Shifting Baseline Syndrome: Causes, Consequences, and Implications." *Frontiers in Ecology and the Environment* 16, no. 4 (2018): 222–30.

Spribille, Toby, Veera Tuovinen, Philipp Reol, Dan Vanderpool, Heimo Wolinski, M. Catherine Aime, Kevin Schneider, et al. "Basidiomycete Yeasts in the Cortex of Ascomycete Macrolichens." *Science (New York, NY)* 353, no. 6298 (2016): 488–92.

Stafford, Barbara Maria. *Voyage into Substance: Art, Science, Nature, and the Illustrated Travel Account, 1760–1840*. Cambridge: MIT Press, 1984.

Stettler, Reinhard F. *Cottonwood and the River of Time: On Trees, Evolution and Society*. Seattle: University of Washington Press, 2009.

Stokes, Dale. *The Fish in the Forest: Salmon and the Web of Life*. Berkeley: University of California Press, 2014.

Tanselle, G. Thomas. "A Rationale of Collecting." *Studies in Bibliography* 51 (1998): 1–25.

Thiers, Barbara. *Herbarium: The Quest to Preserve and Classify the World's Plants*. Portland, OR: Timber Press, 2020.

Thoreau, Henry David. *Walden and Civil Disobedience*. New York: Signet, 1989.

Timler, Kelsey, and Dancing Water Sandy. "Gardening in Ashes: The Possibilities and Limitations of Gardening to Support Indigenous Health and Well-Being in the Context of Wildfires and Colonialism." *International Journal of Environmental Research and Public Health* 17, no. 9 (2020): 3273.

Tortell, Philippe D. "Earth 2020: Science, Society, and Sustainability in the Anthropocene." *Proceedings of the National Academy of Sciences* 117, no. 16 (2020): 8683–91.

Tuan, Yi-fu. *Space and Place: The Perspective of Experience*. Minneapolis : University of Minnesota Press, 1977.

Wagamese, Richard. *One Native Life*. Madeira Park, BC: Douglas and MacIntyre, 2008.

Wandersee, James H., and Elisabeth E. Schussler. "Preventing Plant Blindness." *The American Biology Teacher* 61, no. 2 (1999): 82–86.

Wang, Guo-dong, Weiwei Zhai, He-chuan Yang, Ruo-xi Fan, Xue Cao, Li Zhong, Lu Wang, et al. "The Genomics of Selection in Dogs and the Parallel Evolution between Dogs and Humans." *Nature Communications* 4, no. 1860 (2013). https://doi.org/10.1038/ncomms2814.

Waring, R.H., and J.F. Franklin. "Evergreen Coniferous Forests of the Pacific Northwest." *Science* 204, no. 4400 (1979): 1380–86.

White, E.B. *The Letters of E.B. White*. Edited by M. White. New York: Harper Collins, 2006.

Wilson, Edward O. *Biophilia*. Cambridge: Harvard University Press, 1984.

Wulf, Andrea. *The Invention of Nature: Alexander von Humboldt's New Worlds*. New York: Alfred A. Knopf, 2015.

Yamauchi, Willow. *Adult Child of Hippies*. London: Insomniac Press, 2010.

Zeller, Suzanne. *Inventing Canada: Early Victorian Science and the Idea of a Transcontinental Nation*. Montreal: McGill-Queen's University Press, 2009.

Zencey, Eric. "The Rootless Professors." In *Virgin Forest: Meditations on History, Ecology, and Culture*, 60–71. Athens: University of Georgia Press, 1998.

LYN BALDWIN is an award-winning teacher and plant conservation biologist who uses art and science to help mitigate society's extinction of experience with the botanical world. From her home in the sagebrush-steppe and coniferous forest of the South Thompson Valley, Lyn teaches botany at Thompson Rivers University in Kamloops, BC. For more than two decades, she has worked to cultivate care between the people and plants of place by sharing the stories she finds with her illustrated field journal in art galleries and science museums, and within the pages of journals such as *The Goose, Camas, Hamilton Arts and Letters, The Fourth River, Terrain*, and *The Journal of Natural History Education and Experience*.